Collins

AQA GCSE Re[...]

Chemistry

Chemistry

AQA GCSE

Revision Guide

Emma Poole

Contents

HT Higher Tier Content

Contents

HT Higher Tier Content

Contents

HT Higher Tier Content

Contents

HT Higher Tier Content

Review Questions

Recap of KS3 Key Concepts

1. A match is used to light a candle.
The wax melts and then moves up the candle wick where it burns.

   a) What is the change of state when the wax melts?
   Name the state **i)** before the change and **ii)** after the change. [2]

   b) The match contains carbon.

   Write a word equation to sum up the reaction that takes place when carbon burns completely in oxygen.

   carbon + ______________ → ______________ [2]

   c) Candle wax, wood and coal are all types of fuel.

   Which of these statements is true for all fuels?
   Tick **one** box.

   Fuels are substances that can be burned to release energy. ☐

   Fuels are always solid at room temperature. ☐

   Fuels are always black. ☐

   When fuels are burned, they react with carbon dioxide in the air. ☐ [1]

   d) Name the poisonous gas that can be produced when fuels that contain carbon are burned in a limited supply of oxygen. [1]

   e) Coal contains a small amount of the element sulfur.

   What is the chemical symbol for sulfur? [1]

   f) Name the gas produced when sulfur burns. [1]

2. Potassium nitrate has the formula KNO_3.

   a) How many different elements are shown in this formula? [1]

   b) Which element is represented by the symbol K? [1]

   c) What is the total number of atoms shown in the formula for potassium nitrate? [1]

3. A student uses a pH meter to measure the pH of four solutions.

   a) Tick the correct box to show whether each of the solutions tested is **acidic**, **neutral** or **alkaline**.

Sample	pH	Acidic	Neutral	Alkaline
A	12			
B	10			
C	7			
D	2			

[4]

b) The student placed the pH meter in a beaker of distilled water after testing each of the solutions.

Why did they do this? [1]

c) The student tested a sample of soil from their garden. The soil was slightly acidic.

What could the student put onto the soil to neutralise it? [1]

4 Edward and his family visit the seaside.
Edward has a bucket that contains a mixture of sand and sea water.

a) How could Edward get a sample of salt from the sea water? [1]

b) How could Edward get a sample of sand from the mixture? [1]

c) Mixtures, compounds and elements are different types of substance.
Sea water is a mixture, sand is a compound and oxygen is an element.

Draw **one** line from each type of substance to link it to its definition.

Type of Substance	Definition
Mixture	Contains atoms of two or more elements, which are chemically joined.
Compound	Contains two or more elements or compounds, which are not chemically joined.
Element	Contains only one type of atom.

[2]

5 Rose placed some blue copper sulfate solution into a test tube.
She added a spatula of silver-coloured iron filings and stirred the mixture.
After 10 minutes the solution was green and the solid was orange-brown.

a) Why did Rose stir the mixture? [1]

b) Suggest a safety precaution that Rose should take during this experiment. [1]

c) How did Rose know that a chemical reaction had taken place? [1]

d) Name the element produced during this reaction. [1]

e) Write a word equation to sum up this reaction. [2]

Total Marks .... / 27

Atoms, Elements, Compounds and Mixtures

You must be able to:

- Define the terms: atom, element, mixture and compound
- Write word equations for reactions
- Describe and explain how mixtures can be separated by physical processes.

Atoms, Elements and Compounds

- All substances are made of **atoms**.
- An atom is the smallest part of an **element** that can exist.
- An element is a substance that contains only one sort of atom.
- There are about 100 different elements.
- Elements are displayed in the periodic table.
- The atoms of each element are represented by a different chemical symbol, e.g. sodium = Na, carbon = C and iron = Fe.
- Most substances are **compounds**.
- A compound contains atoms of two or more elements, which are chemically combined in fixed proportions.
- Compounds are represented by a combination of numbers and chemical symbols called a 'chemical formula'.
- Scientists use chemical formulae to show:
  - the different elements in a compound
  - how many atoms of each element one molecule of the compound contains.
- For example:
  - water, H_2O, contains 2 hydrogen (H) atoms and 1 oxygen (O) atom
  - sulfuric acid, H_2SO_4, contains 2 hydrogen (H) atoms, 1 sulfur (S) atom and 4 oxygen (O) atoms.
- Compounds can only be separated into their component elements by chemical reactions or electrolysis.

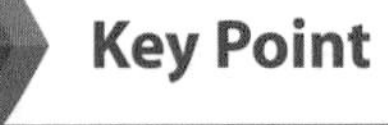

Key Point

Elements are made of only one type of atom. They are displayed in the periodic table and can be represented by one or two letters called symbols.

Equations

- You can sum up what has happened during a chemical reaction by writing a word equation or balanced symbol **equation**.
- The **reactants** (the substances that react) are on the left-hand side of the equation.
- The **products** (the new substances that are formed) are on the right-hand side of the equation.
- The total mass of the products of a chemical reaction is always equal to the total mass of the reactants. This is because no atoms are lost or made.
- The products of a chemical reaction are made from exactly the same atoms as the reactants.
- For example, when magnesium is burned in oxygen, magnesium oxide is produced. This can be written in a word equation:

magnesium + oxygen ⟶ magnesium oxide

Separating Mixtures

- **Mixtures** consist of two or more elements or compounds, which are not chemically combined.
- The components of a mixture retain their own properties, e.g. in a mixture of iron and sulfur, the iron is still magnetic and the sulfur is still yellow.
- Mixtures can be separated by physical processes – these processes do not involve chemical reactions.
- **Filtration** is used to separate soluble solids from insoluble solids, e.g. a mixture of salt (soluble) and sand (insoluble) can be separated by dissolving the salt in water and then filtering the mixture.
- **Crystallisation** is used to obtain a soluble solid from a solution, e.g. salt crystals can be obtained from a solution of salty water:
  1. The mixture is gently warmed.
  2. The water evaporates leaving crystals of pure salt.
- **Simple distillation** is used to obtain a solvent from a solution.

REQUIRED PRACTICAL

Analysis and purification of water samples from different sources.

Sample Method	Hazards and Risks
1. Use a pH probe or suitable indicator to analyse the pH of the sample. 2. Set up the equipment as shown. 3. Heat a set volume to 100°C so that the water changes from liquid to gas. 4. The water collects in the condenser and changes state from gas to liquid. Collect this pure water in a beaker. 5. When all the water from the sample has evaporated, measure the mass of solid that remains to find the amount of dissolved solids present in the sample.	• There is a risk of the experimenter burning themselves on hot equipment, so care must be taken during and after the heating process.

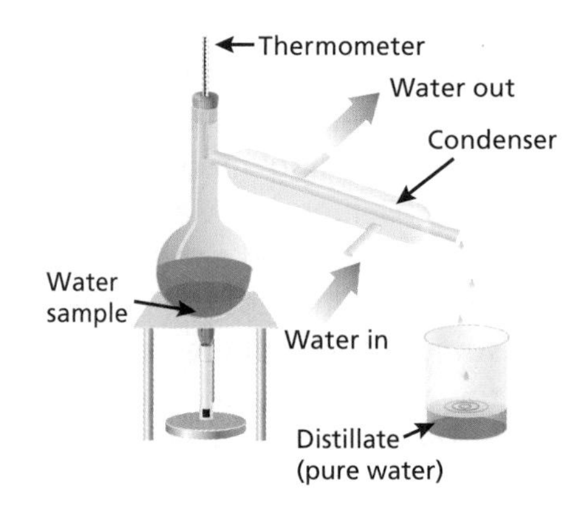

- **Fractional distillation** is used to separate mixtures in which the components have different boiling points, e.g. oxygen and nitrogen can be obtained from liquid air by fractional distillation because they have different boiling points.
- **Chromatography** is used to separate the different soluble, coloured components of a mixture, e.g. the different colours added to a fizzy drink can be separated by chromatography.

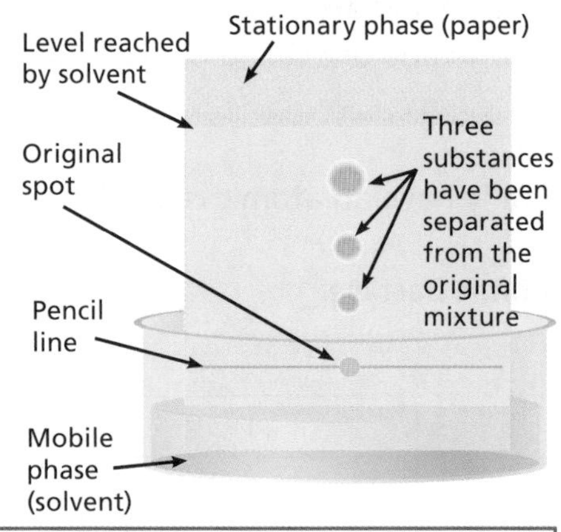

Quick Test

1. Define the term 'atom'.
2. How is an element different to a compound?
3. What does the formula $CaCO_3$ tell you about the compound calcium carbonate?
4. What process can be used to extract pure water from salt water?

Key Words

atom
element
compound
equation
reactants
products
mixture
filtration
crystallisation
simple distillation
fractional distillation
chromatography

Atoms and the Periodic Table

You must be able to:

- Explain how the scientific model of the atom has changed over time
- Recall the relative electrical charge and mass of subatomic particles
- Use the atomic number and mass number to work out the number of protons, neutrons and electrons in an atom or ion
- Describe what an isotope is
- Work out the electron configuration of the first 20 elements.

Scientific Models of the Atom

- In early models, atoms were thought to be tiny spheres that could not be divided into simpler particles.
- In 1898, Thomson discovered electrons and the representation of the atom had to be changed.
- Overall, an atom is neutral, i.e. it has no charge.
- Thomson thought atoms contained tiny, negative **electrons** surrounded by a sea of positive charge. This was the 'plum-pudding' model.
- Later, Geiger and Marsden carried out an experiment in which they bombarded a thin sheet of gold with alpha particles.
- Although most of the positively charged alpha particles passed straight through the atoms, a tiny number were deflected back towards the source.
- Rutherford looked at these results and concluded that the positive charge in an atom must be concentrated in a very small area.
- This area was named the 'nucleus' and the resulting model became known as the 'nuclear' model of the atom.
- Bohr deduced that electrons must orbit the nucleus at specific distances, otherwise they would spiral inwards.
- Later experiments showed that the nucleus is made of smaller particles:
  - some of which have a positive charge and are called **protons**
  - some of which have a neutral charge and are called **neutrons**.

Key Point

Scientists look at the evidence available and use it to put together a model of what appears to be happening. As new evidence emerges, they re-evaluate the model. If the model no longer works, they change it.

The Geiger and Marsden Experiment

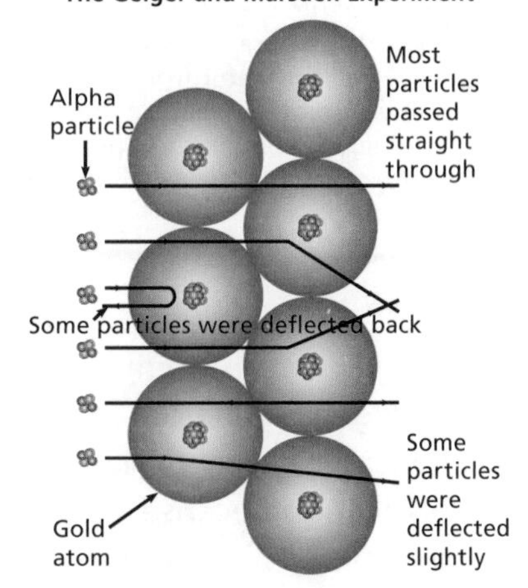

Subatomic Particles

- Atoms are very, very small and typically have an atomic radius of about 0.1nm or 1×10^{-10}m.
- Atoms contain three types of subatomic particle:

Subatomic Particle	Relative Mass	Relative Charge
proton	1	+1
neutron	1	0
electron	very small	–1

- Almost all of the mass of an atom is in the nucleus.
- However, the radius of the nucleus is less than $\frac{1}{10\,000}$ of the atomic radius of the atom, so most of an atom is empty space.
- Atoms have no overall charge because they contain an equal number of protons and electrons.
- All atoms of a particular element have the same number of protons.

The Nuclear Model

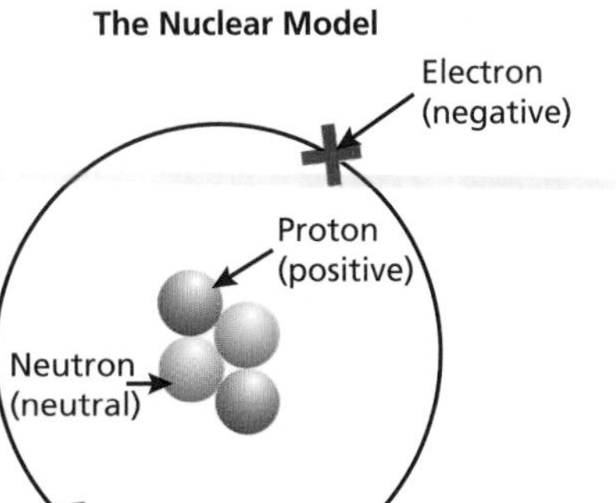

- Atoms of different elements have different numbers of protons.
- The number of protons in an atom is called its **atomic number**.
- The sum of the protons *and* neutrons in an atom is its **mass number**.
- In the modern periodic table, elements are arranged in order of increasing atomic number.

number of neutrons = mass number – atomic number

Key Point

A nanometer (nm) is one billionth of a metre, i.e. $\frac{1}{1\,000\,000\,000}$ or 1×10^{-9}.

How many protons, electrons and neutrons are there in $^{23}_{11}Na$?
11 protons
11 electrons
12 neutrons (23 – 11)

23 is the mass number and 11 is the atomic number.

Isotopes and Ions

- **Isotopes** of an element have the same number of protons but a different number of neutrons, i.e. they have the same atomic number but a different mass number.
- For example, chlorine has two isotopes:

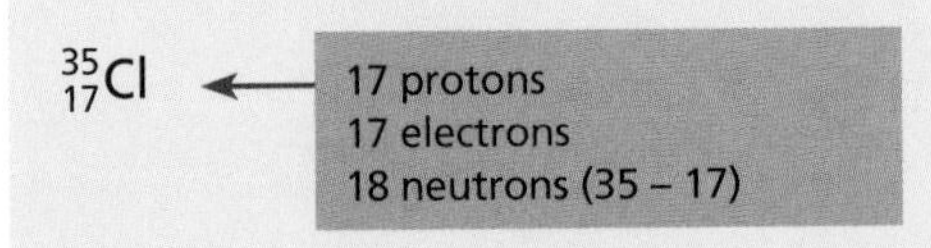

$^{37}_{17}Cl$

17 protons
17 electrons
20 neutrons (37 – 17)

- Atoms can gain or lose electrons to become **ions**:
  - Metal atoms lose electrons to form positive ions.
  - Non-metal atoms gain electrons to form negative ions.

$^{23}_{11}Na^{+}$

Sodium is a metal. It forms positive ions when it loses an electron.

11 protons
10 electrons (11 – 1)
12 neutrons

$^{19}_{9}F^{-}$

Fluorine is a non-metal. It forms negative ions when it gains an electron.

9 protons
10 electrons (9 + 1)
10 neutrons

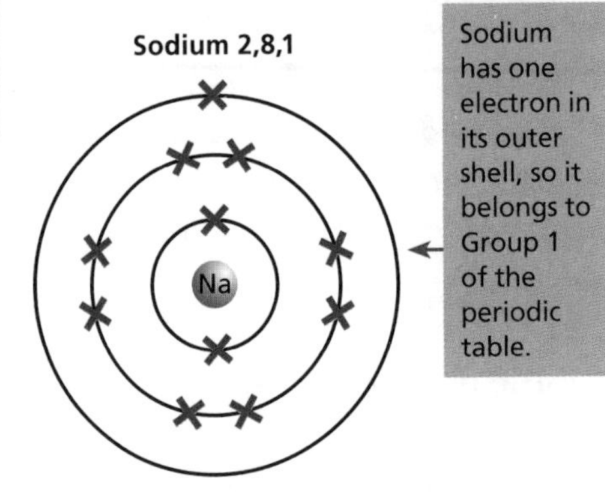

Sodium has one electron in its outer shell, so it belongs to Group 1 of the periodic table.

Electron Configuration

- The electrons in an atom occupy the lowest available shell or energy level.
- For the first 20 elements:
  - the first shell can only hold a maximum of two electrons
  - the next two shells can each hold a maximum of eight electrons.
- The **electron configuration** of an atom shows how the electrons are arranged around the nucleus in shells.

Quick Test

1. Describe the plum-pudding model of the atom.
2. An atom of potassium has an atomic number of 19 and a mass number of 39. State the number of protons, neutrons and electrons in this atom.
3. An ion of potassium-39 has a 1+ charge. State the number of protons, neutrons and electrons in this ion.

Key Words

electron
proton
neutron
atomic number
mass number
isotope
ion
electron configuration

The Periodic Table

You must be able to:

- Describe the key stages in the development of the periodic table
- Describe and explain the properties of Group 0, Group 1 and Group 7 elements
- Describe and explain the properties of the transition metals.

The Development of the Periodic Table

- When John Newlands tried to put together a periodic table in 1864, only 63 elements were known. Many were still undiscovered.
- Newlands arranged the known elements in order of atomic weight.
- He noticed periodicity (repeated patterns), although the missing elements caused problems.
- However, strictly following the order of atomic weight created issues – it meant some of the elements were in the wrong place.
- Dimitri **Mendeleev** realised that some elements had yet to be discovered. When he created his table, in 1869, he left gaps to allow for their discovery. He also reordered some elements.
- Each element was placed in a vertical column or 'group' with elements that had similar properties.
- Mendeleev used his periodic table to predict the existence and properties of new elements.
- When the subatomic particles were later discovered, it revealed that Mendeleev had organised the elements in order of increasing atomic number (number of protons).

Key Point

Elements that have the same number of electrons in their outer shell have similar properties.

Key Point

In the periodic table, elements with the same number of electrons in their outer shell are in the same group, e.g. Group 1 elements all have one electron in their outer shell.

Group 0

- The elements in Group 0 are known as the **noble gases**.
- Noble gas atoms have a full outer shell of electrons.
- This means they have a very stable electron configuration, making them very unreactive non-metals.

Group 1

- The elements in Group 1 are known as the **alkali metals**. They:
  - have one electron in their outermost shell
  - have low melting and boiling points that decrease down the group
  - become more reactive down the group.
- This is because the outer electron gets further away from the influence of the nucleus, so it can be lost more easily.
- Alkali metals are stored under oil because they react very vigorously with oxygen and water, including moisture in the air.
- When alkali metals react with water a metal hydroxide is formed and hydrogen gas is given off.

potassium + water → potassium hydroxide + hydrogen

$$2K(s) + 2H_2O(l) \rightarrow 2KOH(aq) + H_2(g)$$

Key Point

Group 1 metals react with oxygen to form metal oxides. For example, sodium reacts with oxygen to form sodium oxide.

- Group 1 metals have a low density – lithium, sodium and potassium are less dense than water, so float on top of it.
- When a metal hydroxide (e.g. potassium hydroxide) is dissolved in water, an alkaline solution is produced.
- Alkali metals react with non-metals to form ionic compounds.
- When this happens, the metal atom loses one electron to form a metal ion with a positive charge (+1).

sodium + chlorine ⟶ sodium chloride
$2Na(s) + Cl_2(g) \longrightarrow 2NaCl(s)$

Alkali Metals Reacting with Water

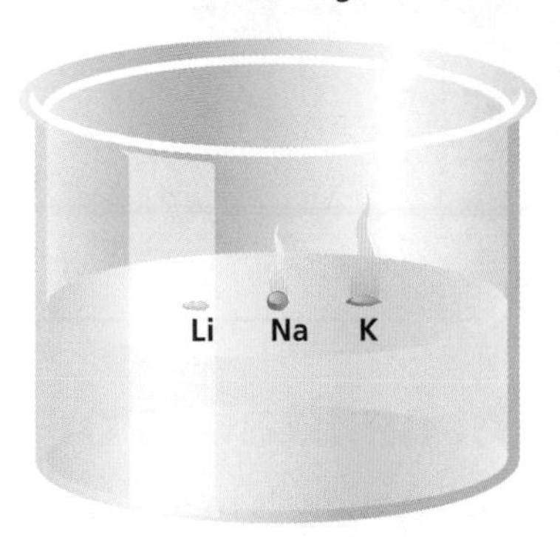

Sodium chloride is a white solid that dissolves in water to form a colourless solution.

Group 7

- The Group 7 elements are non-metals and are known as the **halogens**. They have seven electrons in their outermost shell.
- Reactivity decreases down the group because the outer shell gets further away from the nucleus, so it is less easy to gain an electron.
- Halogens react with metals to produce ionic salts.
- When this happens, the halogen atom gains one electron to form a halide ion with a negative charge (–1).

chlorine + potassium ⟶ potassium chloride
$Cl_2(g) + 2K(s) \longrightarrow 2KCl(s)$

Key Point

The reactivity of Group 1 metals increases down the group as the outer electron is lost more easily.

- A more reactive halogen will **displace** a less reactive halogen from an aqueous solution of its salt. For example:
  - chlorine will displace bromine from potassium bromide and iodine from potassium iodide
  - bromine will displace iodine from potassium iodide.

chlorine + potassium bromide ⟶ potassium chloride + bromine
$Cl_2(g) + 2KBr(aq) \longrightarrow 2KCl(aq) + Br_2(l)$

Key Point

The reactivity of the Group 7 non-metals decreases down the group as it becomes less easy to gain an electron.

The Transition Metals

- The **transition metals** (or transition elements) are in the centre of the periodic table, between Groups 2 and 3.
- They form coloured compounds.
- They have ions with different charges, e.g. Fe^{2+} and Fe^{3+}.
- They can be used as **catalysts** to speed up chemical reactions.
- Like all other metals, the transition metals are good conductors of heat and electricity and can be easily bent or hammered into shape.

Key Point

Transition metals are stronger, harder, have higher melting points and are less reactive than the alkali metals.

Quick Test

1. Name the Russian chemist who designed the periodic table.
2. Why are Group 0 elements unreactive?
3. Why are the alkali metals stored in oil?
4. Give **three** properties of the transition metals.

Key Words

Mendeleev
noble gases
alkali metals
halogens
displace
transition metals
catalyst

States of Matter

You must be able to:

- Recall the meaning of the state symbols in equations
- Describe how the particles move in solids, liquids and gases
- Use the particle model to explain how the particles are arranged in the three states of matter
- HT Describe the limitations of the particle model.

Three States of Matter

- Everything is made of **matter**.
- There are three states of matter: solid, liquid and gas.
- These three states of matter are described by a simple model called the '**particle** model'.
- In this model, the particles are represented by small solid spheres.
- The model can be used to explain how the particles are arranged and how they move in solids, liquids and gases.
- In solids, the particles:
  - have a regular arrangement
  - are very close together
  - vibrate about fixed positions.
- In liquids, the particles:
  - have a random arrangement
  - are close together
  - flow around each other.
- In gases, the particles:
  - have a random arrangement
  - are much further apart
  - move very quickly in all directions.

Solid

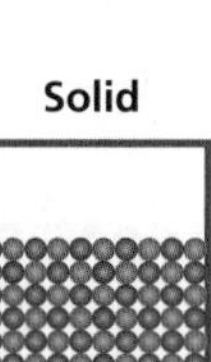

Liquid

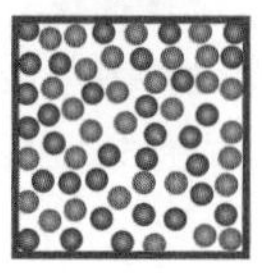

Gas

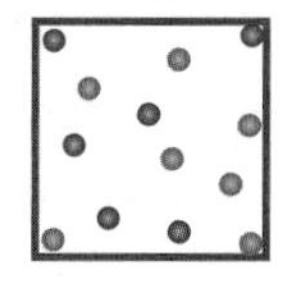

Changing States

- When a substance changes state, e.g. from solid to liquid:
  - the particles themselves stay the same
  - the way the particles are arranged changes
  - the way the particles move changes.
- A pure substance will:
  - melt and freeze at one specific temperature – the **melting point**
  - boil and condense at one specific temperature – the **boiling point**.
- The amount of energy required for a substance to change state depends on the amount of energy required to overcome the forces of attraction between the particles.
- The stronger the forces of attraction:
  - the greater the amount of energy needed to overcome them
  - the higher the melting point and boiling point will be.
- Substances that have high melting points due to strong bonds include ionic compounds, metals and giant covalent structures.

HT **Key Point**

This particle model does have some limitations. It does not take into account:

- the forces between the particles
- the volume (although small) of the particles
- the space between particles.

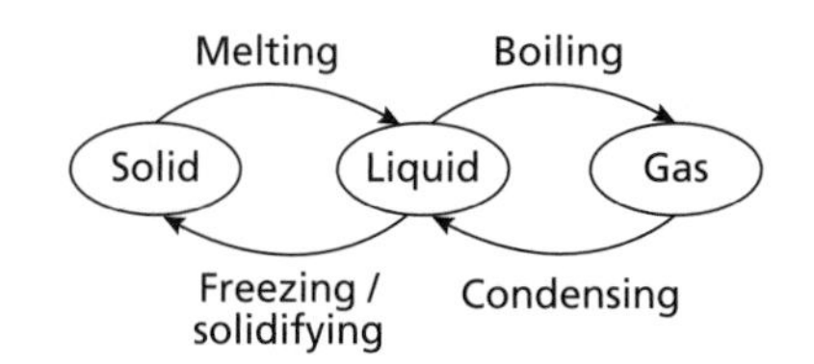

- In substances that contain simple molecules:
  - the bonds within the molecules are strong covalent bonds
  - the forces of attraction between the molecules are much weaker
  - only a little energy is needed to overcome the forces between the molecules, so the melting and boiling points are relatively low.

Key Point

The stronger the forces between particles, the higher the melting point of the substance.

Identifying the State of a Substance

- The melting point and boiling point of a substance can be used to identify its state at a given temperature.

The table below shows the melting points and boiling points of some Group 7 elements.

What is the state of each element at 25°C (room temperature)?

Element	Melting Point (°C)	Boiling Point (°C)
fluorine	–220	–188
chlorine	–102	–34
bromine	–7	59
iodine	114	184

Fluorine and chlorine are gases, bromine is a liquid and iodine is a solid.

25°C is above the boiling points of fluorine and chlorine, so they will be gases.

25°C is above the melting point but below the boiling point of bromine, so it will be a liquid.

25°C is below the melting point of iodine, so it will be a solid.

State Symbols

- Chemical equations are used to sum up what happens in reactions.
- State symbols show the state of each substance involved.

State Symbol	State of Substance
(s)	solid
(l)	liquid
(g)	gas
(aq)	**aqueous** (dissolved in water)

- For example, when solid magnesium ribbon is added to an aqueous solution of hydrochloric acid:
  - a chemical reaction takes place
  - a solution of magnesium chloride is produced
  - hydrogen gas is produced.
- This can be summed up in a symbol equation:

$$Mg(s) + 2HCl(aq) \longrightarrow MgCl_2(aq) + H_2(g)$$

Quick Test

1. What does the state symbol (l) indicate?
2. What does the state symbol (aq) indicate?
3. How do the particles in a gas move?
4. Why do ionic compounds have high melting points?
5. HT State **three** limitations of the particle model.

Key Words

matter
particle
melting point
boiling point
aqueous

Ionic Compounds

You must be able to:

- Describe what an ionic bond is
- Explain how ionic bonding involves the transfer of electrons from metal atoms to non-metal atoms to form ions
- Relate the properties of ionic compounds to their structures.

Chemical Bonds

- There are three types of strong chemical bonds:
  - **ionic bonds**
  - covalent bonds
  - metallic bonds.
- Atoms that have gained or lost electrons are called **ions**.
- Ionic bonds occur between positive and negative ions.

Ionic Bonding

- Ions are formed when atoms gain or lose electrons, giving them an overall charge.
- Ions have a complete outer shell of electrons (the same electronic structure as a noble gas).
- Ionic bonding involves a transfer of electrons from metal atoms to non-metal atoms.
- The metal atoms lose electrons to become positively charged ions.
- The non-metal atoms gain electrons to become negatively charged ions.
- The ionic bond is a strong **electrostatic** force of attraction between the positive metal ion and the negative non-metal ion.

Sodium forms an ionic compound with chlorine.

Describe what happens when two atoms of sodium react with one molecule of chlorine.
Give your answer in terms of electron transfer.

- Sodium belongs to Group 1 of the periodic table. It has one electron in its outer shell.
- Chlorine belongs to Group 7 of the periodic table. It has seven electrons in its outer shell.
- One chlorine molecule contains two chlorine atoms.
- Each sodium atom transfers one electron to one of the chlorine atoms.
- All four atoms now have eight electrons in their outer shell.
- The atoms become ions, Na^+ and Cl^-.
- The compound formed is sodium chloride, NaCl.

$$2Na + Cl_2 \rightarrow 2NaCl$$

Key Point

The arrangement of electrons in an atom can be described in terms of shells or energy levels. Electron configuration diagrams are a good example of using diagrams to represent information.

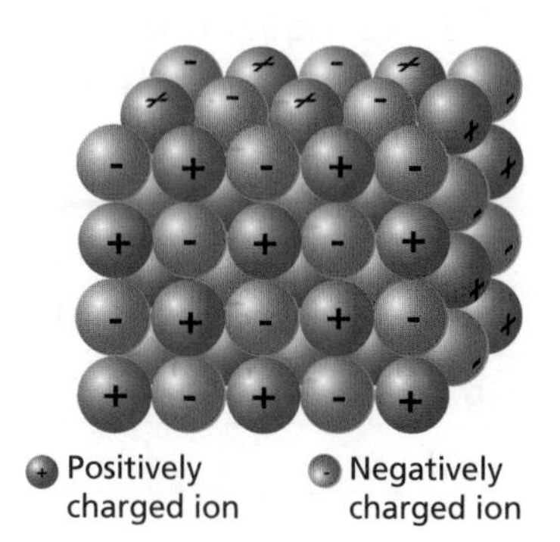

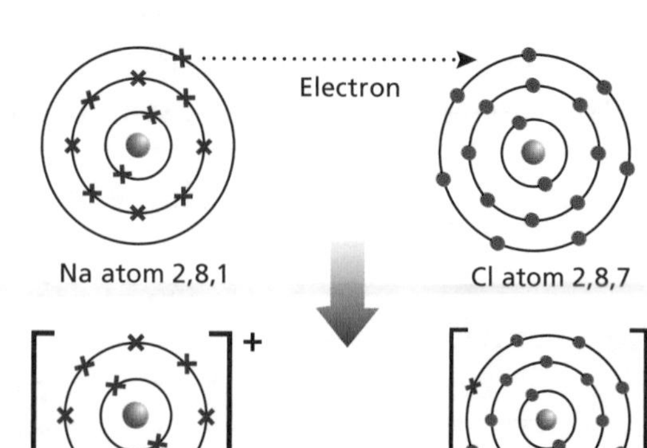

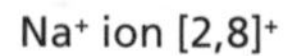

When magnesium is burned, it forms an ionic compound with oxygen.

Describe what happens when two atoms of magnesium react with one molecule of oxygen. Give your answer in terms of electron transfer.

- Magnesium is in Group 2 of the periodic table. It has two electrons in its outer shell.
- Oxygen is in Group 6 of the periodic table. It has six electrons in its outer shell.
- One oxygen molecule contains two oxygen atoms.
- Each magnesium atom loses two electrons to an oxygen atom.
- All four atoms now have eight electrons in their outer shell.
- The atoms become ions, Mg^{2+} and O^{2-}.
- The compound formed is magnesium oxide, MgO.

$$2Mg + O_2 \rightarrow 2MgO$$

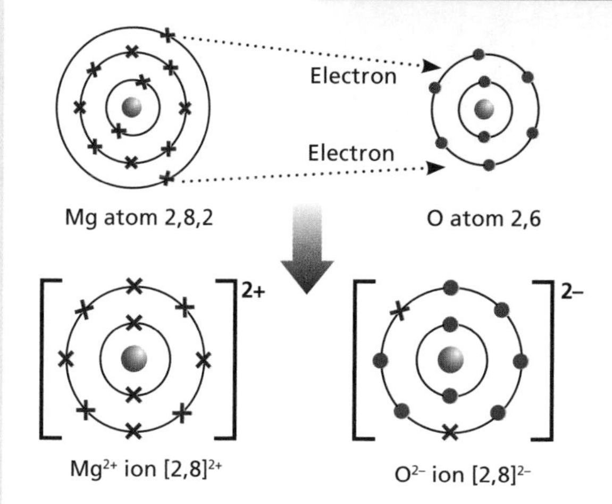

Key Point

An ionic bond is the attraction between oppositely charged ions.

Properties of Ionic Compounds

- Ionic compounds are giant structures of ions.
- They are held together by strong forces of attraction (electrostatic forces) that act in all directions between oppositely charged ions, i.e. ionic compounds are held together by strong ionic bonds.
- Ionic compounds:
  - have high melting and boiling points
  - do *not* conduct electricity when solid, because the ions cannot move
  - do conduct electricity when **molten** or in solution, because the charged ions are free to move about and carry their charge.

Key Point

Ionic compounds have high melting and boiling points because ionic bonds are very strong and it requires lots of energy to overcome them.

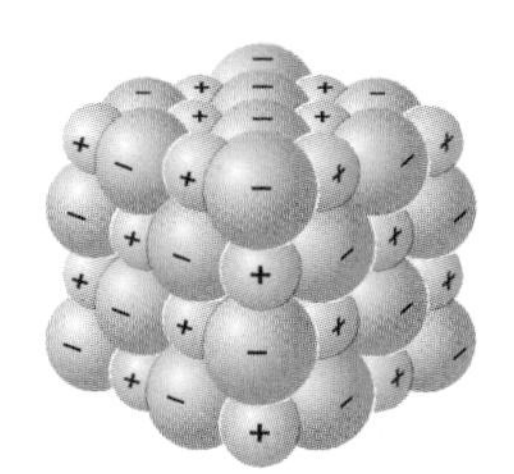

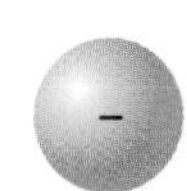

Negatively charged chloride ions

Positively charged sodium ions

Quick Test

1. What is an ion?
2. Why do metals form positively charged ions?
3. The ionic compound potassium chloride contains potassium ions (K^+) and chloride ions (Cl^-). What is the formula of potassium chloride?
4. Why do ionic compounds conduct electricity when molten?

Key Words

ionic bond
ion
electrostatic
molten

Metals

You must be able to:

- Describe when and why metallic bonding occurs
- Explain why metals conduct electricity
- Describe and explain the properties of pure metals and alloys.

Metallic Bonding

- Metallic bonding occurs in:
  - metallic elements, such as iron and copper
  - alloys, such as stainless steel.
- Metals have a giant structure in which electrons in the outer shell are **delocalised** (not bound to one atom).
- This produces a regular arrangement (lattice) of positive ions held together by **electrostatic** attraction to the delocalised electrons.
- A **metallic bond** is the attraction between the positive ions and the delocalised negatively charged electrons.

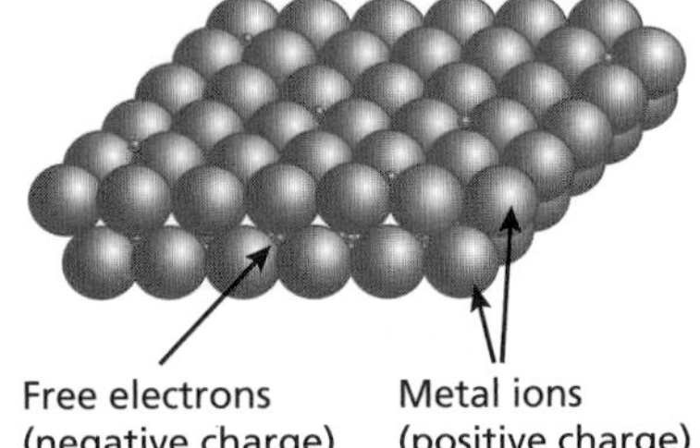

Properties of Metals

- The properties of metals make them very useful.
- Metallic bonds are very strong and most metals have high melting and boiling points. This means that they are useful structural materials.
- The delocalised electrons can move around freely and transfer energy. This makes metals good thermal and electrical conductors.
- The particles in **pure** metals have a regular arrangement.
- The layers are able to slide over each other quite easily, which means that metals can be bent and shaped.
- Traditionally, copper is used to make water pipes because:
  - it is an unreactive metal, so it does not react with water
  - it can be easily shaped.

Metal	Uses	Property
Aluminium	High-voltage power cables, furniture, drinks cans, foil food wrap	Corrosion resistant, **ductile**, **malleable**, good conductivity, low density
Copper	Electrical wiring, water pipes, saucepans	Ductile, malleable, good conductivity
Gold	Jewellery, electrical junctions	Ductile, shiny, good conductivity

Key Point

A metallic bond is the attraction between positive ions and delocalised electrons.

Alloys

- Most metal objects are made from **alloys** – mixtures that contain a metal and at least one other element.
- Pure metals are too soft for many uses.
- In alloys, the added element disturbs the regular arrangement of the metal atoms so the layers do not slide over each other so easily.
- This means alloys are usually stronger and harder than pure metals.

Steel

- Steel is a very useful alloy made from iron.
- Iron oxide can be reduced (see page 38) in a blast furnace to produce iron.
- Molten iron obtained from a blast furnace contains roughly 96% iron and 4% impurities, including carbon, phosphorus and silica.
- Because it is impure, the iron is very **brittle** and has limited uses.
- To produce pure iron, all the impurities have to be removed.
- The atoms in pure iron are arranged in layers that can slide over each other easily, making it soft and malleable.
- Pure iron can be easily shaped, but it is too soft for many practical uses.
- The properties of iron can be changed by mixing it with small quantities of carbon or other metals to make steel.
- The majority of iron is converted into steel.
- Alloys are developed to have the required properties for a specific purpose.
- In steel, the amount of carbon and / or other elements determines its properties:
  - Steel with a high carbon content is hard and strong.
  - Steel with a low carbon content is soft and easily shaped.
  - Stainless steel contains chromium and nickel and is hard and resistant to **corrosion**.

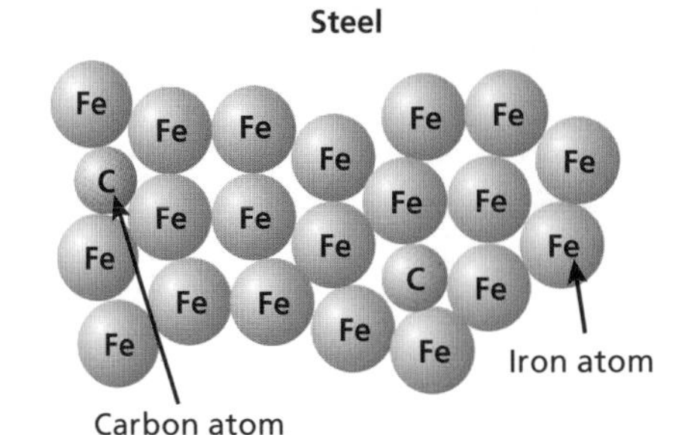

HT Other Useful Alloys

- Pure copper, gold and aluminium are too soft for many uses.
- They are mixed with small amounts of similar metals to make them harder for items in everyday use, e.g. coins.
- Gold is mixed with silver, copper and zinc to form an alloy.
- The carat system shows the amount of gold in the alloy:
  - 24 carat gold is 100% gold
  - 18 carat gold is $\frac{18}{24} \times 100\% = 75\%$ gold.
- Aluminium alloys combine low density with high strength and are used to make aeroplanes.
- Bronze is an alloy of copper and tin.
- It has a bright gold colour and is used to make statues and decorative objects.
- Brass is an alloy of copper and zinc that is hard-wearing and very resistant to corrosion.
- It is used to make water taps and door fittings.

Quick Test

1. Describe what a metallic bond is.
2. Why is copper a good material for water pipes?
3. What is an alloy?
4. Why are alloys more useful than pure metals?
5. Give **two** useful properties that stainless steel has but pure iron does not have.

Key Words

delocalised
electrostatic
metallic bond
pure
ductile
malleable
alloy
brittle
corrosion

Covalent Compounds

You must be able to:

- Describe a covalent bond
- Describe the structure of simple molecules
- Explain the properties of simple molecules
- Describe the giant covalent structures: diamond, graphite and silicon dioxide
- Explain the properties of giant covalent structures.

Covalent Bonding

- A **covalent bond** is a shared pair of electrons between atoms.
- Covalent bonds occur in:
  - non-metallic elements, e.g. oxygen, O_2
  - compounds of non-metals, e.g. sulfur dioxide, SO_2.
- For example, a chlorine atom has seven electrons in its outer shell. In order to bond with another chlorine atom:
  - an electron from each atom is shared
  - this gives each chlorine atom eight electrons in the outer shell
  - each atom now has a complete outer shell.
- Covalent bonds in molecules can be shown using dot and cross diagrams.
- Covalent bonds are very strong.
- Some covalently bonded substances have simple structures:

A Chlorine Molecule (One Covalent Bond)

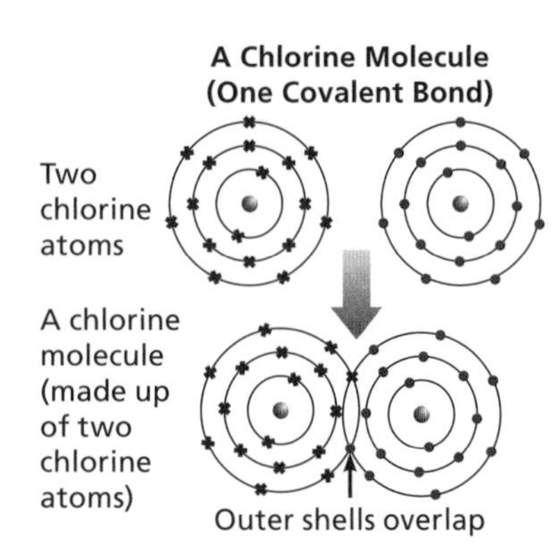

A Molecule of Ammonia (Three Covalent Bonds)

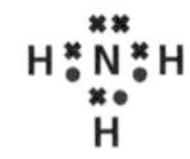

Molecule	Water H_2O	Chlorine Cl_2	Hydrogen H_2	Hydrogen chloride, HCl	Methane CH_4	Oxygen O_2
Method 1	H O H	Cl Cl	H H	H Cl	H C H H H	O O
Method 2	H–O–H	Cl–Cl	H–H	H–Cl	H–C–H (with H above and below C)	O=O (a double bond)

- Others have giant covalent structures, e.g. diamond and silicon dioxide.

Simple Molecules

- Simple molecules contain a relatively small number of non-metal atoms joined together by covalent bonds.
- The molecules have no overall electrical charge, so they cannot conduct electricity.
- Substances that consist of simple molecules usually have low melting and boiling points.
- This is because they have weak **intermolecular** forces (forces of attraction between the molecules).
- These intermolecular forces are very weak compared to the strength of the covalent bonds in the molecules themselves.

Key Point

Simple molecular substances, such as water, have low melting and boiling points. This is because there are only weak forces of attraction between the molecules, which are easily overcome.

- The larger the molecules are, the stronger the intermolecular forces between the molecules become.
- This means that larger molecules have higher melting and boiling points.
- Going down Group 7 of the periodic table, the molecules get larger and their melting and boiling points increase.
- This is demonstrated by their states at room temperature:
  - Fluorine and chlorine are gases.
  - Bromine is a liquid.
  - Iodine is a solid.

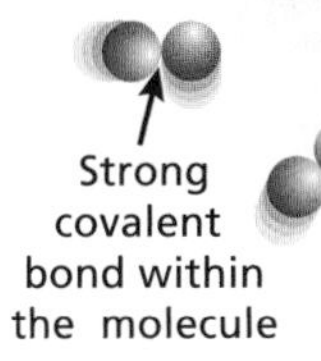

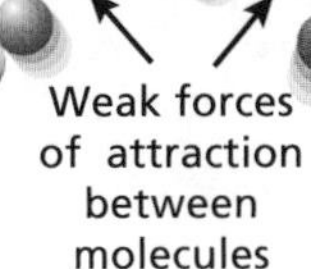

Giant Covalent Structures

- All the atoms in giant covalent structures are linked by strong covalent bonds.
- These bonds must be broken for the substance to melt or boil.
- This means that giant covalent structures have very high melting and boiling points.
- **Diamond** is a form of carbon:
  - It has a giant, rigid covalent structure (lattice).
  - Each carbon atom forms four strong covalent bonds with other carbon atoms.
  - All the strong covalent bonds mean that it is a very hard substance with a very high melting point.
  - There are no charged particles, so it does not conduct electricity.
- **Graphite** is another form of carbon:
  - It also has a giant covalent structure and a very high melting point.
  - Each carbon atom forms three covalent bonds with other carbon atoms.
  - This results in a layered, hexagonal structure.
  - The layers are held together by weak intermolecular forces.
  - This means that the layers can slide past each other, making graphite soft and slippery.
  - One electron from each carbon atom in graphite is **delocalised**.
  - These delocalised electrons allow graphite to conduct heat and electricity.
- **Silicon dioxide** (or **silica**, SiO_2) has a lattice structure similar to diamond:
  - Each oxygen atom is joined to two silicon atoms.
  - Each silicon atom is joined to four oxygen atoms.

Diamond

Covalent bond between carbon atoms

Carbon atom

Graphite

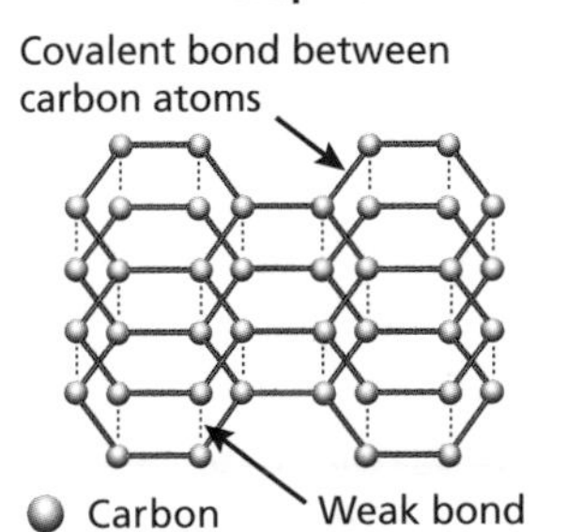

Quick Test

1. What is a covalent bond?
2. Why does hydrogen chloride (HCl) have a low boiling point?
3. How does the size of a simple molecule affect the strength of the intermolecular forces between molecules?
4. Describe the structure of diamond.
5. Explain how graphite can conduct electricity.

Key Words

covalent bond
intermolecular
diamond
graphite
delocalised
silicon dioxide (silica)

Special Materials

You must be able to:

- Recall the structure and uses of graphene and fullerenes
- Recall the size of nanoparticles, fine particles and coarse particles
- Understand the bonding within and between polymer molecules
- Understand why nanoparticles have special properties.

Graphene

- **Graphene** is a form of carbon. It is a single layer of graphite (see page 21).
- The atoms are arranged in a hexagonal structure, just one atom thick.
- Graphene has some special properties.
- It is very strong, a good thermal and electrical conductor and nearly **transparent**.

Fullerenes

- Carbon can also form molecules known as **fullerenes**, which contain different numbers of carbon atoms.
- Fullerene molecules have hollow shapes, including tubes, balls and cages.
- The first fullerene to be discovered was buckminsterfullerene, C_{60}:
  - It consists of 60 carbon atoms.
  - The atoms are joined together in a series of hexagons and pentagons.
  - It is the most symmetrical and, therefore, most stable fullerene.
- Carbon nanotubes are cylindrical fullerenes with some very useful properties.
- Fullerenes can be used:
  - to deliver drugs in the body
  - in lubricants
  - as catalysts
  - for reinforcing materials, e.g. the frames of tennis rackets, so that they are strong but still lightweight.

Structure of Buckminsterfullerene

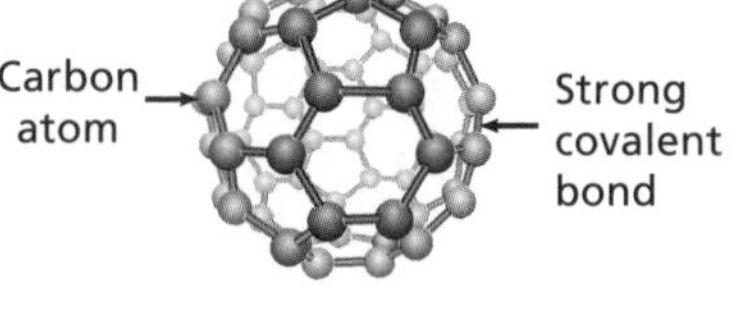

Structure of a Nanotube

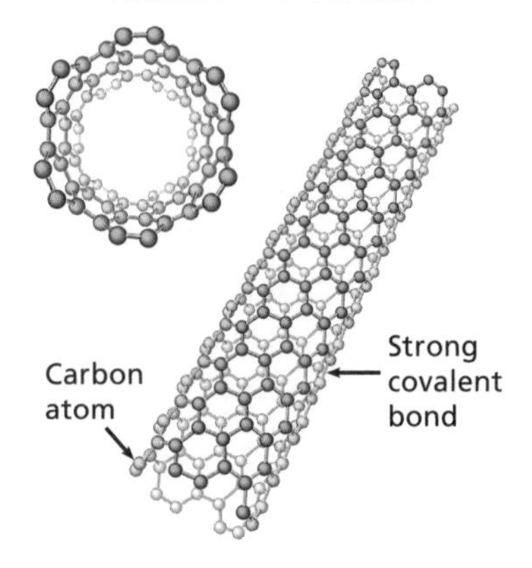

Polymers

- Polymers consist of very large molecules.
- Plastics are synthetic (man-made) polymers.
- The atoms within the polymer molecules are held together by strong covalent bonds.
- The intermolecular forces between the large polymer molecules are also quite strong.
- This means that polymers are solid at room temperature.
- Poly(ethene), commonly known as polythene, is produced when lots of ethene molecules are joined together in an addition polymerisation reaction (see pages 70–71).
- It is cheap and strong and is used to make plastic bottles and bags.

$$n\,CH_2{=}CH_2 \rightarrow \left(-CH_2-CH_2- \right)_n$$

Sizes of Particles and their Properties

- Coarse particles (often called 'dust' by scientists) have a diameter between 1×10^{-5}m and 2.5×10^{-6}m.
- Fine particles have a diameter between 100nm and 2500nm or 1×10^{-7}m and 2.5×10^{-6}m.
- **Nanoparticles** have a diameter between 1nm and 100nm or 1×10^{-9}m and 1×10^{-7}m.
- Nanoscience is the study of these very small structures.
- Small particles have a high surface area to volume ratio.
- Changing the size of particles has a dramatic effect on this ratio.
- For example, if the length of side of a cube decreases by a factor of 10:
  - the surface area decreases by $10 \times 10 = 100$
  - the volume decreases by $10 \times 10 \times 10 = 1000$
  - the surface area to volume ratio increases tenfold.
- This is important for catalysts, as having a large surface area improves their effectiveness.

Key Point

One nanometre is 0.000 000 001m (one billionth of a metre) and is written as 1nm or 1×10^{-9}m.

Nanoparticles

- Nanoparticles contain only a few hundred atoms.
- They can combine to form structures called nanostructures.
- Nanostructures can be manipulated, so materials can be developed that have new and specific properties.
- The properties of nanoparticles are different to the properties of the same materials in bulk, e.g. nanoparticles are more sensitive to light, heat and magnetism.
- In nanoparticles, the atoms can be placed into exactly the right position so smaller quantities are needed to achieve the required properties / effects.
- Nanoparticles are used in sun creams.
- They provide better skin coverage and, therefore, more effective protection from the sun's harmful ultraviolet rays.
- However, concerns remain that these nanoparticles are so small they could get into and damage human cells or cause problems in the environment.
- Research into nanoparticles is leading to the development of:
  - new drug delivery systems
  - synthetic skin for burn victims
  - computers and technology
  - catalysts for fuel cells
  - stronger and lighter construction materials
  - new cosmetics and deodorants
  - fabrics that prevent the growth of bacteria.

Key Point

The use of nanoparticles is a good example of a science-based technology, for which the hazards have to be considered alongside the benefits. However, because the technology is so new, data is uncertain and incomplete.

Quick Test

1. What is special about the structure of graphene?
2. Which is the most stable fullerene?
3. How big are nanoparticles?
4. What potential problems could nanoparticles cause?

Key Words

graphene
transparent
fullerene
nanoparticles

Practice Questions

Atoms, Elements, Compounds and Mixtures

1. What are the substances that react together in a chemical reaction called?
   Tick **one** box.

   Products ☐ Mixtures ☐ Reactants ☐ Ions ☐ [1]

2. When copper metal is heated it reacts with oxygen to form copper oxide.

   Write a word equation for this reaction. [1]

3. Why is the total mass of the reactants in a chemical reaction always equal to the total mass of the products? [1]

4. Define the term 'mixture'. [1]

5. Suggest a technique that could be used to separate the components of a food colouring used in cupcake icing. [1]

6. What technique could be used to extract a sample of pure salt from a solution of salt and water? [1]

7. What property allows oxygen and nitrogen to be obtained by the fractional distillation of liquid air?
   Tick **one** box.

   They are in different groups of the periodic table ☐

   They have different boiling points ☐

   They are both soluble ☐

   They are both gaseous at room temperature (25°C) ☐ [1]

Total Marks .... / 7

Atoms and the Periodic Table

1. Complete **Table 1** to show the relative mass of the different subatomic particles.

Table 1

Subatomic Particle	Relative Mass
proton	....
....	1
electron	....

[3]

2 An ion of potassium is represented as ${}^{39}_{19}K^+$.

Give the number of protons, electrons and neutrons in this ion of potassium. [3]

3 What is the radius of a typical atom?
Tick **one** box.

1×10^{-12}m ☐

1×10^{-10}m ☐

1×10^{-8}m ☐

1×10^{-3}m ☐ [1]

4 An atom of magnesium is represented by:

${}^{25}_{12}Mg$

a) How many protons does this atom of magnesium have? [1]

b) How many neutrons does this atom of magnesium have? [1]

c) How many electrons does this atom of magnesium have? [1]

5 Define the term 'isotope'. [2]

Total Marks .... / 12

The Periodic Table

1 **a)** Which group in the periodic table do the alkali metals belong to? [1]

b) How many electrons do the alkali metals have in their outermost shell? [1]

c) Explain, in terms of electrons, why the alkali metals get more reactive as you go down the group. [2]

d) Why do the alkali metals have to be stored under oil? [2]

Total Marks .... / 6

Practice Questions

States of Matter

1. State symbols are used to add extra information to equations.

   a) What does the state symbol (g) mean? [1]

   b) What does the state symbol (aq) mean? [1]

2. HT Give **three** limitations of the particle model. [3]

3. Calcium reacts with hydrochloric acid to produce calcium chloride and hydrogen. The equation for the reaction is shown below:

   $Ca(s) + 2HCl(aq) \rightarrow CaCl_2(aq) + H_2(g)$

   What is the state of the:

   a) Calcium? [1]

   b) Hydrochloric acid? [1]

   c) Hydrogen? [1]

Total Marks .... / 8

Ionic Compounds

1. Sodium fluoride is an ionic compound.

   a) Suggest why sodium fluoride has a high melting point. [2]

   b) Does sodium fluoride conduct electricity when solid?
   You must explain your answer. [1]

   c) Does sodium fluoride conduct electricity when molten?
   You must explain your answer. [1]

   d) Does sodium fluoride conduct electricity when dissolved in water to form an aqueous solution?
   You must explain your answer. [1]

2 **Table 1** shows the charge on some metal ions and non-metal ions.

Table 1

Metal Ions	Non-Metal Ions
Sodium, Na^+	Chloride, Cl^-
Magnesium, Mg^{2+}	Oxide, O^{2-}
Potassium, K^+	Fluoride, F^-
Calcium, Ca^{2+}	Sulfide, S^{2-}

a) What is an ion? [1]

b) In terms of electron transfer, explain why chloride ions have a 1– charge. [2]

c) In terms of electron transfer, explain why magnesium ions have a 2+ charge. [2]

d) Use the information in the table to determine the formula of:

i) Potassium chloride [1]

ii) Magnesium sulfide [1]

iii) Calcium oxide. [1]

e) Magnesium oxide has a very high melting point and can be used to line furnaces.

Explain why the compound magnesium oxide has a high melting point. [2]

Total Marks .... / 15

Metals

1 **a)** What is the chemical symbol for gold? [1]

b) Name the type of bonding that occurs in gold. [1]

c) Pure gold is too soft for many uses.

Why is pure gold soft? [2]

d) Gold is often made into an alloy.

What is an 'alloy'? [1]

e) Gold is used to make components for computers because it is a very good electrical conductor.

Why is gold a good electrical conductor? [2]

Total Marks .... / 7

Covalent Compounds

1 **Figure 1** shows the structure of diamond.

Figure 1

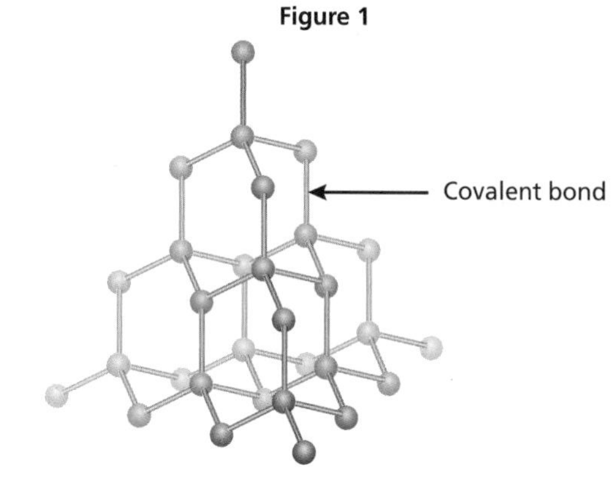

a) Diamond is a form of which element? [1]

b) Why is diamond so hard? [1]

c) Why is diamond a poor electrical conductor? [1]

2 Methane molecules have the formula CH_4.

a) Explain why methane is a poor electrical conductor. [1]

b) Explain why methane has a low boiling point. [2]

3 **Figure 2** shows the outer electrons in a magnesium atom and in a sulfur atom.

Figure 2

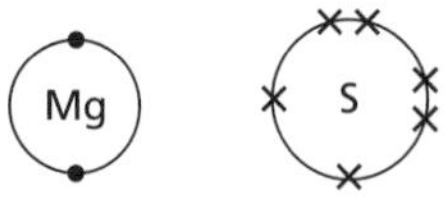

Mg S

a) i) In which group of the periodic table is magnesium? [1]

ii) In which group of the periodic table is sulfur? [1]

iii) Draw a diagram to show the outer electrons in a magnesium ion and in a sulfide ion. [2]

b) The compound magnesium sulfide is a solid at room temperature.

What type of structure does magnesium sulfide have?
Tick **one** box.

Giant metallic	☐	Giant covalent	☐
Simple molecular	☐	Giant ionic	☐

[1]

c) Explain why magnesium sulfide does not conduct electricity when solid. [2]

4 Ammonia, NH_3, and water, H_2O, are both simple molecules.
Both of these compounds contain covalent bonds.

a) Define the term 'covalent bond'. [1]

b) **Figure 3** shows the outer electrons in a nitrogen atom and in a hydrogen atom.

Figure 3

N H

Complete **Figure 4** below to show the electron arrangement in an ammonia molecule. [1]

Figure 4

H N H
H

c) Explain why ammonia and water do **not** conduct electricity. [1]

Total Marks .... / 16

Special Materials

1 a) What size are nanoparticles?
Tick **one** box.

1 to 100nm	☐	1 to 10nm	☐
Less than 1nm	☐	100 to 1000nm	☐

[1]

b) Suggest a possible problem that nanoparticles could cause. [1]

Total Marks .... / 2

Conservation of Mass

You must be able to:

- Understand the law of conservation of mass
- Work out the relative formula mass of substances
- Understand why reactions that involve gases may appear to show a change in mass
- HT Write balanced half equations and ionic equations.

The Conservation of Mass

- In a chemical reaction, the total mass of the products is equal to the total mass of the reactants.
- This idea is called the **conservation of mass**.
- Mass is conserved (kept the same) because no atoms are lost or made.
- Chemical symbol equations must always be balanced to show this, i.e. there must be the same number of atoms of each element on both sides of the equation.
- For example, when solid iron reacts with copper(II) sulfate solution, a reaction takes place, producing solid copper and iron(II) sulfate solution:

$$Fe(s) + CuSO_4(aq) \longrightarrow Cu(s) + FeSO_4(aq)$$

- HT A **half equation** can be used to show what happens to one reactant in a chemical reaction, with electrons written as e^-.
- HT The balanced symbol equation for the reaction between iron and copper(II) sulfate can be split into two half equations:
  - The iron atoms lose two electrons to form Fe^{2+} ions.

$$Fe(s) \longrightarrow Fe^{2+}(aq) + 2e^-$$

  - The Cu^{2+} ions gain two electrons to form copper atoms.

$$Cu^{2+}(aq) + 2e^- \longrightarrow Cu(s)$$

- HT **Ionic equations** can be used to simplify complicated equations.
- HT They just show the species that are involved in the reaction.
- HT The spectator ions (ions not involved in the reaction) are not included.
- HT For example, when silver nitrate solution is added to sodium chloride solution, a white precipitate of silver chloride is produced:

$$AgNO_3(aq) + NaCl(aq) \longrightarrow AgCl(s) + NaNO_3(aq)$$

- HT In this reaction, the nitrate ions and the sodium ions are spectator ions, so the ionic equation is:

$$Ag^+(aq) + Cl^-(aq) \longrightarrow AgCl(s)$$

- HT In chemistry, the term '**species**' refers to the different atoms, molecules or ions that are involved in a reaction.

Key Point

The total mass of the products of a chemical reaction is always equal to the total mass of the reactants. This is because no atoms are lost or made. The products are made from exactly the same atoms as the reactants.

Relative Formula Mass

- The **relative formula mass (M_r)** of a compound is the sum of the **relative atomic masses (A_r)** of all the atoms in the numbers shown in the formula. It does not have a unit.
- The relative atomic masses of the atoms are shown in the periodic table.

What is the relative formula mass of carbon dioxide, CO_2?

Relative formula mass = (12 × 1) + (16 × 2)

= 44

CO_2 contains 1 carbon atom with a relative atomic mass of 12 and 2 oxygen atoms with a relative atomic mass of 16.

Calculate the relative formula mass of calcium nitrate, $Ca(NO_3)_2$.

Relative formula mass = 40 + (14 × 2) + (16 × 6)

= 164

Remember that everything inside a set of brackets is multiplied by the number outside the brackets, so $Ca(NO_3)_2$ contains 1 calcium, 2 nitrogen and 6 oxygen atoms.

- Due to conservation of mass, the sum of the relative formula masses of all the reactants is always equal to the sum of the relative formula masses of all the products.

Key Point

The relative atomic mass is an average value that takes account of the abundance of the isotopes of an element.
25% of chlorine atoms have a mass of 37.
75% of chlorine atoms have a mass of 35.
The relative atomic mass of chlorine =

$(\frac{25}{100} \times 37) + (\frac{75}{100} \times 35)$
= 35.5

Apparent Changes in Mass

- Some reactions appear to involve a change in mass.
- This happens when reactions are carried out in a non-closed system and include a gas that can enter or leave.
- For example, when magnesium is burned in air to produce magnesium oxide, the mass of the solid increases.
- This is because when the magnesium is burned, it combines with oxygen from the air and the oxygen has mass:

$$2Mg(s) + O_2(g) \rightarrow 2MgO(s)$$

- If the mass of oxygen is included, the total mass of all the reactants is equal to the total mass of all the products.
- When calcium carbonate is heated, it decomposes to form calcium oxide and carbon dioxide:

$$CaCO_3(s) \rightarrow CaO(s) + CO_2(g)$$

- The mass of the solid decreases because one of the products is a gas, which escapes into the air.
- If the mass of carbon dioxide is included, the total mass of all the reactants is equal to the total mass of all the products.

Key Point

Some reactions appear to involve a change in mass. This happens when reactions are carried out in a non-closed system, so gases can enter or leave.

Quick Test

1. State the law of conservation of mass.
2. Why must a symbol equation balance?
3. Calculate the relative formula mass of water, H_2O.
4. Calculate the relative formula mass of calcium carbonate, $CaCO_3$.
5. When iron is burned, iron oxide is produced and the mass of the solid increases. Why does the mass of the solid increase during this reaction?

Key Words

conservation of mass
HT **half equation**
HT **ionic equation**
HT **species**
relative formula mass (M_r)
relative atomic mass (A_r)

Quantitative Chemistry

Amount of Substance

You must be able to:

- HT Recall the number of particles in one mole of any substance
- HT Calculate the amount of a substance in moles
- HT Calculate the mass of reactants or products from balanced equations
- HT Calculate the balancing numbers in equations from the masses of the reactants and the products by using moles
- HT Calculate the volume of a given amount of a gas.

Amount of Substance

- A **mole** (mol) is a measure of the number of particles (atoms, ions or molecules) contained in a substance.
- One mole of any substance (element or compound) contains the same number of particles – six hundred thousand billion billion or 6.02×10^{23}.
- This value is known as the **Avogadro constant**.
- The mass of one mole of a substance is its relative atomic mass or relative formula mass in grams.

One mole of sodium atoms contains 6.02×10^{23} atoms.
The relative atomic mass of sodium is 23.0.
One mole of sodium atoms has a mass of 23.0g.

Key Point

One mole of any substance (element or compound) will always contain the same number of particles – six hundred thousand billion billion or 6.02×10^{23}. This value is known as the Avogadro constant.

Calculating the Amount of Substance

- You can calculate the amount of substance (number of moles) in a given mass of a substance using the formula:

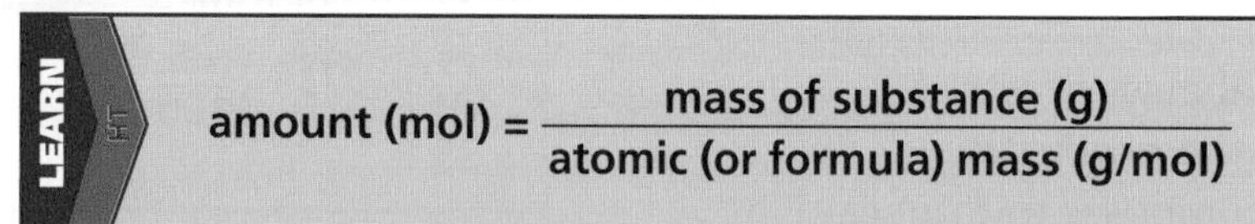

LEARN HT

$$\textbf{amount (mol)} = \frac{\textbf{mass of substance (g)}}{\textbf{atomic (or formula) mass (g/mol)}}$$

Calculate the number of moles of carbon dioxide in 33g of the compound.

$$\text{amount} = \frac{\text{mass of substance}}{\text{formula mass}}$$
$$= \frac{33}{44}$$
$$= 0.75\text{mol}$$

Formula mass = 12 + (16 × 2)
= 44

Balanced Equations

- Balanced equations:
  - show the number of moles of each product and reactant
  - can be used to calculate the mass of the reactants and products.
- The numbers needed to balance an equation can be calculated from the masses of the reactants and the products using moles.

Aluminium oxide can be reduced to produce aluminium:

$$Al_2O_3 \rightarrow 2Al + 1\frac{1}{2}O_2$$

Calculate the mass of aluminium oxide needed to produce 540g of aluminium.

$$\text{amount of aluminium} = \frac{540}{27} = 20\text{mol}$$

$$\text{amount of aluminium oxide required} = \frac{20}{2} = 10\text{mol}$$

$$\text{formula mass of aluminium oxide} = (27 \times 2) + (16 \times 3) = 102$$

$$\text{mass of aluminium oxide needed} = 10 \times 102 = 1020\text{g}$$

The equation shows that one mole of aluminium oxide produces two moles of aluminium.

amount of aluminium = $\frac{\text{mass}}{\text{atomic mass}}$

Relative atomic masses (A_r): Al = 27 and O = 16.

The equation shows that one mole of aluminium oxide is needed to produce two moles of aluminium, so divide by two.

mass of alumnium oxide needed = amount (mol) × formula mass

- The numbers needed to balance an equation can be calculated from the masses of the reactants and the products using moles.

In a chemical reaction, 72g of magnesium was reacted with exactly 48g of oxygen molecules to produce 120g of magnesium oxide.

Use the number of moles of reactants and products to write a balanced equation for the reaction.

$$\text{amount of Mg} = \frac{72}{24} = 3\text{mol}$$

$$\text{amount of } O_2 = \frac{48}{32} = 1.5\text{mol}$$

$$\text{amount of MgO} = \frac{120}{40} = 3\text{mol}$$

$$3Mg + 1.5O_2 \rightarrow 3MgO$$

$$2Mg + O_2 \rightarrow 2MgO$$

Use the masses of the reactants to calculate the number of moles present.

Divide the number of moles of each substance by the smallest number (1.5) to give the simplest whole number ratio.

This shows that 2 moles of magnesium react with 1 mole of oxygen molecules to produce 2 moles of magnesium oxide.

Limiting Reactants

- Sometimes when two chemicals react together, one chemical is completely used up during the reaction.
- When one chemical is used up, it stops the reaction going any further. It is called the **limiting reactant**.
- The other chemical, which is not used up, is said to be in excess.

Moles of a Gas

- At room temperature and pressure, one mole of any gas takes up a volume of $24dm^3$.
- At room temperature and pressure:

volume = amount (mol) × $24dm^3$

Quick Test

1. 69g of sodium reacts with chlorine to produce sodium chloride:
   $$2Na + Cl_2 \rightarrow 2NaCl$$
   a) Calculate the number of moles of sodium present.
   b) Calculate the number of moles of chlorine (Cl_2) that would be required to react exactly with the sodium.
   c) Calculate the mass of chlorine that would be required to react exactly with the sodium.

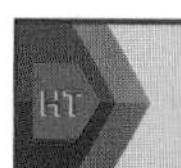

Key Words

mole (mol)
Avogadro constant
limiting reactant

Titration

You must be able to:

- HT Recall the units for calculating the concentration of solutions
- HT Be able to work out the amount of solute in a solution of known volume and concentration
- HT Describe how to carry out a titration using a strong acid and a strong alkali
- HT Work out the concentration of a solution using data from titrations.

Concentration

- The **concentration** of a solution is often measured using units of mol/dm^3.

LEARN HT

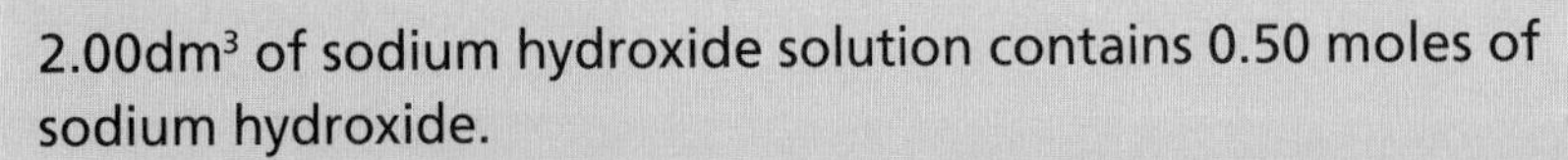

$$\text{concentration of a solution} = \frac{\text{amount of substance (mol)}}{\text{volume (dm}^3\text{)}}$$

- If 1.00 mole of **solute** is dissolved to form a solution that has a volume of $1.00dm^3$, the solution has a concentration of $1.00mol/dm^3$.

Key Point

There are $1000cm^3$ in $1.00dm^3$, so $500cm^3$ has a volume of $0.500dm^3$.

$2.00dm^3$ of sodium hydroxide solution contains 0.50 moles of sodium hydroxide.

Work out the concentration of the solution.

$$\text{concentration of a solution} = \frac{\text{amount of substance (mol)}}{\text{volume (dm}^3\text{)}}$$

$$= \frac{0.50mol}{2.00dm^3} = 0.25mol/dm^3$$

Substitute the values into the formula.

- Occasionally concentrations are expressed in g/dm^3.
- If 10.0g solute is dissolved to form a solution that has a volume of $1.00dm^3$, the solution has a concentration of $10.0g/dm^3$.

Key Point

The concentration of a solution is found by dividing the amount of substance (in moles or grams) by the volume (in dm^3).

Carrying Out a Titration

- Acids and alkalis react together to form a neutral solution.
- **Titration** is an accurate technique that can be used to find out how much of an acid is needed to neutralise an alkali.
- When neutralisation takes place, the hydrogen ions (H^+) from the acid join with the hydroxide ions (OH^-) from the alkali to form water (neutral pH).

$$H^+(aq) + OH^-(aq) \longrightarrow H_2O(l)$$

- You must use a suitable **indicator** in titrations.
- If you have a strong acid and strong alkali, you could use methyl orange or phenolphthalein.
- Hydrochloric acid, nitric acid and sulfuric acid are all strong acids.
- Aqueous sodium hydroxide and aqueous potassium hydroxide are strong alkalis.

Key Point

Indicators are one colour in acids and another colour in alkalis. They are used to show the end point of the titration.

REQUIRED PRACTICAL	
Determination of the reacting volumes of solutions of a strong acid and a strong alkali by titration.	
Sample Method **1.** Wash and rinse a pipette with the alkali being used. **2.** Use the pipette to measure out a known and accurate volume of the alkali. **3.** Place the alkali in a clean, dry conical flask. **4.** Add a suitable indicator, e.g. phenolphthalein. **5.** Place the flask on a white tile so the colour can be seen clearly. **6.** Place the acid in a burette that has been carefully washed and rinsed with the acid. **7.** Take a reading of the volume of acid in the burette (initial reading). **8.** Carefully add the acid to the alkali, swirling the flask to thoroughly mix. **9.** Continue until the indicator just changes colour. This is called the end point. **10.** Take a reading of the volume of acid in the burette (final reading). **11.** Calculate the volume of acid added (i.e. subtract the initial reading from the final reading).	**Hazards and Risks** • Acids and alkalis can damage the skin or eyes, so eye protection must be worn and any spillages wiped up.

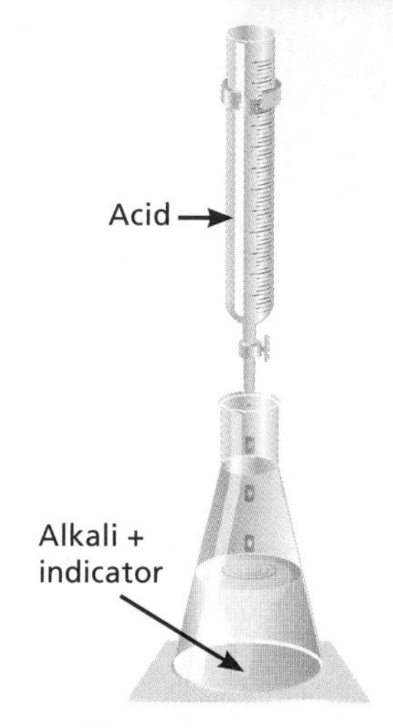

- Titration can be used to find the concentration of an acid or alkali, providing the following are known:
  - the relative volumes of acid and alkali used
  - the concentration of the other acid or alkali.
- Break down the calculation:
  1. Write down a balanced equation for the reaction to determine the ratio of moles of acid to alkali involved.
  2. Calculate the number of moles in the solution of known volume and concentration. You can work out the number of moles in the other solution from the balanced equation.
  3. Calculate the concentration of the other solution.

A titration is carried out and $0.04dm^3$ hydrochloric acid neutralises $0.08dm^3$ sodium hydroxide of concentration $1.00mol/dm^3$.

Calculate the concentration of the hydrochloric acid.

$$HCl + NaOH \rightarrow NaCl + H_2O$$

number of moles of NaOH = volume × concentration

$$= 0.08dm^3 \times 1.00mol/dm^3 = 0.08mol$$

$$\text{concentration of HCl} = \frac{\text{number of moles of HCl}}{\text{volume of HCl}}$$

$$= \frac{0.08mol}{0.04dm^3} = 2.00mol/dm^3$$

Write the balanced symbol equation for the reaction.

The balanced equation shows that the amount of hydrochloric acid is equal to the amount of sodium hydroxide, i.e. 0.08mol.

HT Key Point

This method can be repeated to check results and can then be performed without an indicator in order to obtain the salt.

HT Quick Test

1. 1.50 moles of solute is dissolved in $1.00dm^3$ of solution. What is the concentration of the solution?
2. 2.00g of solute is dissolved in $2.00dm^3$ of solution. What is the concentration of the solution?
3. 0.20 moles of solute is dissolved in $500cm^3$ of solution. What is the concentration of the solution?

Key Words

concentration
solute
titration
indicator

Percentage Yield and Atom Economy

You must be able to:

- Explain why a reaction may not produce the theoretical yield of the product
- Calculate the percentage yield for a reaction
- Calculate the atom economy of a reaction
- HT Explain why a particular reaction pathway may be chosen, using appropriate information.

Percentage Yield

- Atoms are never lost or gained in a chemical reaction.
- However, it is *not* always possible to obtain the calculated amount of product:
  - If the reaction is reversible, it might not go to completion.
  - Some product could be lost when it is separated from the reaction mixture.
  - Some of the reactants may react in different ways to the expected reaction.
- The amount of product obtained is called the yield.
- The **percentage yield** can be calculated using the formula:

LEARN

$$\textbf{percentage yield} = \frac{\textbf{yield from reaction}}{\textbf{maximum theoretical yield}} \times 100$$

Key Point

Percentage yield is used to compare the actual yield obtained from a reaction with the maximum theoretical yield.

A percentage yield of 100% would mean that no product had been lost.

Calculating Yield

Calculate how much calcium oxide can be produced from 50.0kg of calcium carbonate.

$CaCO_3 \rightarrow CaO + CO_2$

$[40 + 12 + (3 \times 16)] \rightarrow [40 + 16] + [12 + (2 \times 16)]$

$100 \rightarrow 56 + 44$

100 : 56

100kg of $CaCO_3$ produces 56kg of CaO

So, 1kg of $CaCO_3$ produces $\frac{56}{100} = 0.56$kg of CaO

And, 50kg of $CaCO_3$ produces $0.56 \times 50 = 28$kg of CaO

Relative atomic masses (A_r): Ca = 40, C = 12, O = 16

Write down the equation.

Work out the M_r of each substance.

Check that the total mass of reactants equals the total mass of products. If they are not the same, check your work.

The question only mentions calcium oxide and calcium carbonate, so you can now ignore the carbon dioxide. You just need the ratio of mass of reactant to mass of product.

Use the ratio to calculate how much calcium oxide can be produced.

50kg of calcium carbonate ($CaCO_3$) is expected to produce 28kg of calcium oxide (CaO).
A company heats 50kg of calcium carbonate in a kiln and obtains 22kg of calcium oxide. Calculate the percentage yield.

percentage yield = $\frac{22}{28} \times 100 = 78.6\%$

Atom Economy

- **Atom economy** is a measure of the amount of reactant that ends up in a useful product.
- Scientists try to choose reaction pathways that have a high atom economy.

- This is important for economic reasons and for **sustainable development**, as more products are made and less waste is produced.
- The percentage atom economy is calculated using the formula:

$$\text{atom economy} = \frac{\text{relative formula mass of the desired product}}{\text{sum of the relative formula mass of all the products/reactants}} \times 100$$

> **Key Point**
>
> Sustainable development meets the needs of the current generation without compromising the ability of future generations to meet their own needs.

The Production of Ethanol

- Ethanol can be produced in two different ways: **hydration** and **fermentation**.
- During hydration, ethene is reacted with steam to form ethanol:

$$C_2H_4 + H_2O \longrightarrow C_2H_5OH$$

- The atom economy for the hydration method is 100%.
- The hydration of ethene is an **addition reaction** – all the reactant atoms end up in the desired product.
- Ethanol can also be produced by the fermentation of glucose:

$$C_6H_{12}O_6 \longrightarrow 2C_2H_5OH + 2CO_2$$

atom economy = $\frac{92}{180}$ x 100 = 51.1%

- The atom economy for this reaction pathway is much lower – only about half of the atoms in the reactants end up in the desired product.

HT Choosing a Reaction Pathway

- Comparing the atom economy of two competing reaction pathways is important.
- However, it is just one of the factors that scientists have to consider when they choose which method to use.
- Important factors to consider when choosing a reaction pathway include:
  - the atom economy
  - cost of reactants
  - the percentage yield
  - the rate of reaction (see pages 60–61)
  - the equilibrium position (see pages 62–63)
  - the usefulness of by-products.

> **Key Point**
>
> Although the hydration of ethene has a high atom economy, ethene is produced from crude oil, which is non-renewable, and the process requires high temperatures, which can be expensive to maintain.

Quick Test

1. Why might the actual yield of a reaction be less than the theoretical yield of the reaction?
2. A reaction has a theoretical yield of 13g but an actual yield of 8.5g. What is the percentage yield of this reaction?
3. Why are reactions that have a high atom economy good for the environment?
4. The equation for the formation of ammonia, NH_3, from its elements is shown below. What is the percentage atom economy of this reaction?
   $N_2 + 3H_2 \rightarrow 2NH_3$

Key Words

percentage yield
atom economy
sustainable development
hydration
fermentation
addition reaction

Reactivity of Metals

You must be able to:

- Recall that when metals form metal oxides, the metals are oxidised
- Recall that metals can be placed into a reactivity series
- Recall that a more reactive metal can displace a less reactive metal from a solution of its salt
- Explain how metals less reactive than carbon can be extracted from their oxides
- HT Explain oxidation and reduction in terms of electron transfer.

Oxidation and Reduction

- In **oxidation** reactions, a substance often gains oxygen.
- In **reduction** reactions, a substance often loses oxygen.
- Oxidation and reduction always occur together.
- Metals react with oxygen to form metal oxides.
- For example, when magnesium is burned in air it reacts with oxygen to form magnesium oxide.
- The magnesium gains oxygen in the reaction, so it is oxidised:

magnesium + oxygen → magnesium oxide

$$2Mg + O_2 \rightarrow 2MgO$$

- Metal oxides can be reduced by removing oxygen.
- For example, when lead(IV) oxide is heated with carbon:
  - the lead(IV) oxide loses oxygen so it is reduced
  - the carbon gains oxygen so it is oxidised.

lead(IV) oxide + carbon → lead + carbon dioxide

$$PbO_2 + C \rightarrow Pb + CO_2$$

Key Point

Oxidation and reduction always occur together.

The Reactivity Series

- When metals react, their atoms lose electrons to form positive metal ions.
- Some metals lose electrons more easily than others.
- The more easily a metal atom loses electrons, the more reactive it is.
- The reaction of metals with acid and water can be used to place them in order of reactivity. This is called the **reactivity series**.
- Metals react with acids to produce metal salts and hydrogen.
- Lithium, sodium and potassium are very reactive metals – they react vigorously with water to produce a metal hydroxide solution and hydrogen.
- These metals are placed at the top of the reactivity series.
- Lithium, sodium and potassium would react so vigorously with dilute acids that it would not be safe to carry out the reactions.
- Calcium, magnesium, zinc and iron are fairly reactive metals – they react quickly with acids and slowly with water.
- Very unreactive metals, like copper and gold, do not react with acids or water and are placed at the bottom of the periodic table.
- Reactivity series often include carbon and hydrogen for comparison.

Potassium, K
Sodium, Na
Calcium, Ca
Magnesium, Mg
Aluminium, Al
Carbon, C
Zinc, Zn
Iron, Fe
Tin, Sn
Lead, Pb
Hydrogen, H
Copper, Cu
Silver, Ag
Gold, Au
Platinum, Pt

↑ Most reactive

Key Point

Some metals lose electrons more easily than others. The more easily a metal atom loses electrons, the more reactive it is.

Displacement Reactions

- In a **displacement reaction** a more reactive metal will displace a less reactive metal from a solution of its salt.
- Magnesium is more reactive than copper, so magnesium will displace copper from a solution of copper sulfate:

magnesium + copper sulfate → copper + magnesium sulfate

$$Mg(s) + CuSO_4(aq) \rightarrow Cu(s) + MgSO_4(aq)$$

Extraction of Metals

- The method of **extraction** of a metal depends on how reactive it is.
- Unreactive metals (e.g. gold) exist as elements at the Earth's surface.
- However, most metals are found as metal oxides, or as compounds that can be easily changed into metal oxides.
- Metals that are less reactive than carbon (e.g. iron and lead) can be extracted from their oxides by heating with carbon:

iron oxide + carbon → iron + carbon dioxide

- The iron oxide loses oxygen, so it is reduced.
- The carbon gains oxygen, so it is oxidised.

Key Point

Metals that are more reactive than carbon (e.g. aluminium) are extracted from molten compounds by electrolysis.

HT Losing or Gaining Electrons

- Not all reduction and oxidation reactions involve oxygen.
- Because of this, scientists use the following rules:
  - Oxidation is the loss of electrons.
  - Reduction is the gain of electrons.
- The balanced symbol equation for the reaction between magnesium and oxygen can be split into two ionic equations:

$$2Mg \rightarrow 2Mg^{2+} + 4e^-$$
$$O_2 + 4e^- \rightarrow 2O^{2-}$$

The magnesium atoms lose electrons to become magnesium ions – the magnesium is oxidised. The oxygen atoms gain electrons to become oxide ions – the oxygen is reduced.

Quick Test

1. calcium + oxygen → calcium oxide
   Which substance is oxidised in this reaction?
2. Complete the word equation:
   iron + copper sulfate → ________ + ________
3. By what method should lead be extracted from lead oxide?
4.  Calcium reacts with hydrochloric acid to form calcium chloride and hydrogen. The equation for this reaction can be split into two ionic equations:
   $Ca \rightarrow Ca^{2+} + 2e^-$
   $2H^+ + 2e^- \rightarrow H_2$
   a) Which species is oxidised? Explain your answer.
   b) Which species is reduced? Explain your answer.

Key Words

oxidation
reduction
reactivity series
displacement reaction
extraction

The pH Scale and Salts

You must be able to:

- Use the pH scale to show how acidic or alkaline a solution is
- Explain that in neutralisation reactions H^+ ions react with OH^- ions to produce water, H_2O
- Recall how soluble salts can be made from soluble and insoluble bases
- HT Describe the difference between a weak acid and a strong acid.

The pH Scale

- When substances dissolve in water, they **dissociate** into their individual ions:
  - Hydroxide ions, $OH^-(aq)$, make solutions alkaline.
  - Hydrogen ions, $H^+(aq)$, make solutions acidic.
- The pH scale is a measure of the acidity or alkalinity of an **aqueous** solution:
  - A solution with a pH of 7 is neutral.
  - Aqueous solutions with a pH less than 7 are acidic.
  - The closer to a pH of zero, the stronger the acid.
  - Aqueous solutions with a pH of more than 7 are alkaline.
  - The closer to a pH of 14, the stronger the alkali.
- The pH of a solution can be measured using a pH probe or universal indicator.
- **Indicators** are dyes that change colour depending on whether they are in acidic or alkaline solutions:
  - Litmus changes colour from red to blue or vice versa.
  - Universal indicator is a mixture of dyes that shows a range of colours to indicate how acidic or alkaline a substance is.

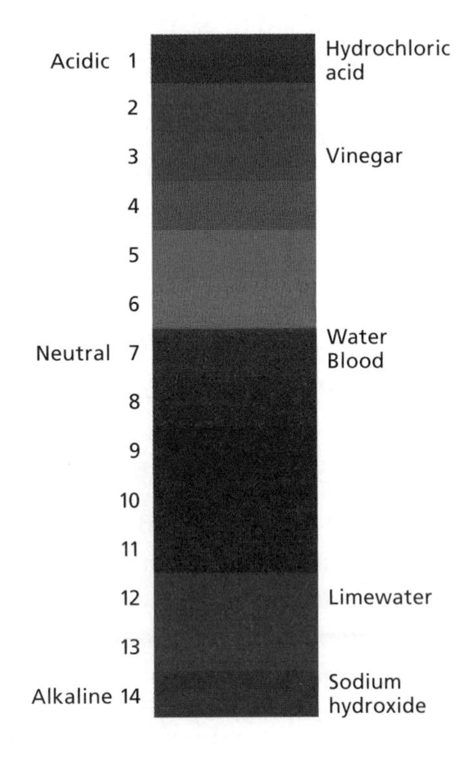

Neutralisation of Acids

- Soluble bases are called alkalis.
- Acids are neutralised by bases.

acid + metal hydroxide → salt + water

- Acids contain hydrogen ions, $H^+(aq)$.
- Alkalis contain hydroxide ions, $OH^-(aq)$.
- When an acid reacts with an alkali, the H^+ and OH^- ions react together to produce water, H_2O, which has a pH of 7.

$$H^+(aq) + OH^-(aq) \rightarrow H_2O(l)$$

- This type of reaction is called **neutralisation** because:
  - acid is neutralised by an alkali
  - the solution that remains has a pH of 7, showing it is neutral.
- Acids can also be neutralised by metal oxides and metal carbonates:

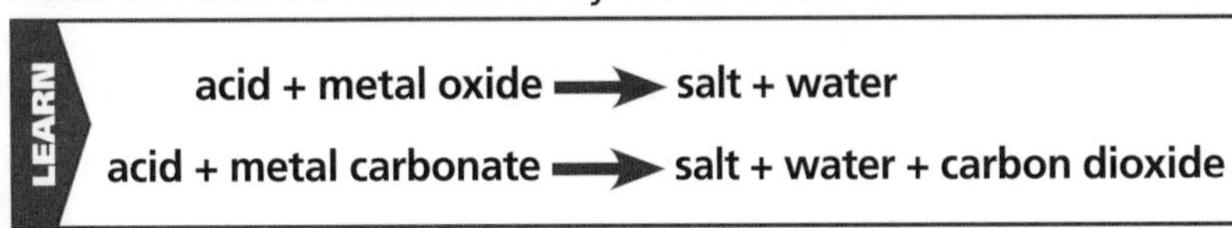

- A salt is produced when the hydrogen in the acid is replaced by a metal ion.
- The name of the salt produced depends on the acid used:
  - Hydrochloric acid produces chloride salts.
  - Nitric acid produces nitrate salts.
  - Sulfuric acid produces sulfate salts.

Soluble Salts from Insoluble Bases

- Soluble salts can be made by reacting acids with insoluble bases, such as metal oxides, metal hydroxides and metal carbonates.

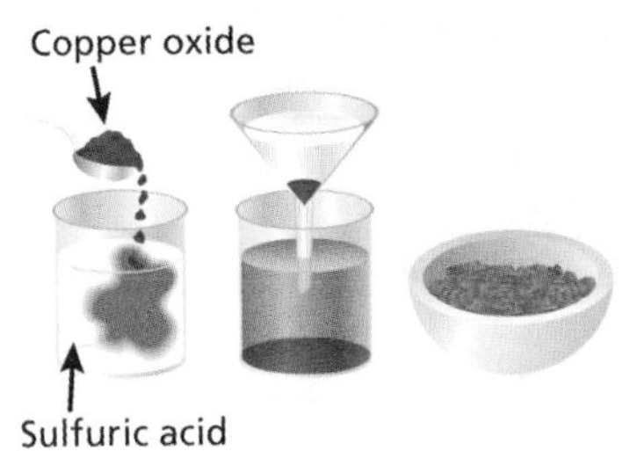

Add copper oxide to sulfuric acid → Filter to remove any unreacted copper oxide → Evaporate using a water bath or electric heater to leave behind blue crystals of the 'salt' copper sulfate

REQUIRED PRACTICAL

Preparation of a pure, dry sample of a soluble salt from an insoluble oxide or carbonate.

Sample Method	Hazards and Risks
1. Add the metal oxide or carbonate to a warm solution of acid until no more will react. 2. Filter the excess metal oxide or carbonate to leave a solution of the salt. 3. Gently warm the salt solution so that the water evaporates and crystals of salt are formed.	• Corrosive acid can cause damage to eyes, so eye protection must be used. • Hot equipment can cause burns, so care must be taken when the salt solution is warmed.

HT Strong and Weak Acids

- **Strong acids** are completely **ionised** (split up into ions) in water.
- Hydrochloric acid is a strong acid:

$$HCl(g) + aq \longrightarrow H^{+}(aq) + Cl^{-}(aq)$$

- Ethanoic acid is a **weak acid**:

$$CH_3COOH(l) + aq \rightleftharpoons CH_3COO^{-}(aq) + H^{+}(aq)$$

- The pH of a solution is a measure of the concentration of H^+ ions.
- A pH decrease of one unit indicates that the concentration of hydrogen ions has increased by a factor of 10.
- For a given concentration of acid, a strong acid will have a higher concentration of hydrogen ions and, therefore, a lower pH.
- The terms 'dilute' and 'concentrated' are also applied to acids sometimes.
- An acid that has a concentration of $2mol/dm^3$ is more concentrated than an acid that has a concentration of $0.5mol/dm^3$.

The 'aq' in the equation indicates water.

Notice how the sign for a reversible reaction is used in the equation.

HT Key Point

Strong acids such as hydrochloric acid, nitric acid and sulfuric acid are completely ionised in water.

Weak acids such as ethanoic acid, citric acid and carbonic acid are only partially ionised in water.

Quick Test

1. What is the pH of a neutral solution?
2. Complete the general equation below:
   acid + alkali → ________ + ________
3. HT What is a strong acid?
4. HT The pH of a solution changes from 6 to 4. What happens to the concentration of hydrogen ions?

Key Words

dissociate
aqueous
indicator
neutralisation
HT **strong acid**
HT **ionised**
HT **weak acid**

Electrolysis

You must be able to:

- Explain why ionic compounds conduct electricity when molten or in aqueous solution
- Predict the products of the electrolysis of simple ionic compounds and explain how electrolysis can be used to extract reactive metals
- HT Write half equations for the reactions that take place at the electrodes during electrolysis.

Electrolysis

- **Electrolysis** is the use of an electrical current to break down compounds containing ions into their constituent elements.
- The substance being broken down is called the **electrolyte**.
- The **electrodes** are made from solids that conduct electricity.
- During electrolysis:
  - negatively charged ions move to the **anode** (positive electrode)
  - positively charged ions move to the **cathode** (negative electrode).
- Electrolysis can be used to separate ionic compounds into elements.
- For example, lead bromide can be split into lead and bromine:
  - The lead bromide is heated until it melts.
  - The positively charged lead ions move to the negative electrode (cathode).
  - Here they gain electrons to form lead atoms – pure lead is produced at this electrode.
  - The negatively charged bromide ions move to the positive electrode (anode).
  - Here they lose electrons to form bromine atoms, which join together to form bromine molecules – bromine is released at this electrode.

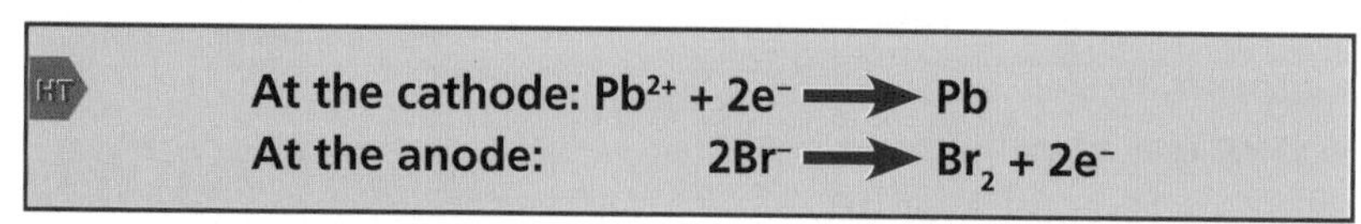

HT At the cathode: $Pb^{2+} + 2e^- \longrightarrow Pb$

At the anode: $2Br^- \longrightarrow Br_2 + 2e^-$

- Ionic substances can only conduct electricity when they are molten or dissolved in water.

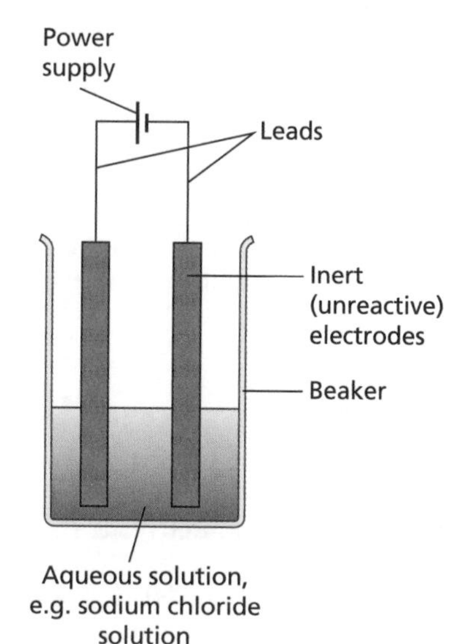

Key Point

For the electrolysis of molten ionic compounds, the electrodes used must be inert so that they do not react with the electrolyte or the products.

HT Oxidation and Reduction

- Reduction occurs when positively charged ions gain electrons at the negative electrode.
- Oxidation occurs when negatively charged ions lose electrons at the positive electrode.
- In a redox reaction both reduction and oxidation occur.
- Reactions that take place at the electrodes during electrolysis can be represented by half-equations.
- For example, in the electrolysis of molten copper chloride:
  - Copper is deposited at the negative electrode.

$$Cu^{2+} + 2e^- \longrightarrow Cu$$

The copper ions gain electrons so they are reduced.

Key Point

You can remember this by thinking of the word OILRIG:

- Oxidation Is Loss of electrons (OIL)
- Reduction Is Gain of electrons (RIG).

- Chlorine gas is given off at the positive electrode.

$$2Cl^- \longrightarrow Cl_2 + 2e^-$$

The chloride ions lose electrons so they are oxidised.

Remember that chlorine exists as molecules.

Extraction of Metals

- Metals that are more reactive than carbon can be extracted from their ores using electrolysis.
- Electrolysis requires lots of heat and electrical energy, making it an expensive process.
- Aluminium is obtained by the electrolysis of aluminium oxide that has been mixed with **cryolite** (a compound of aluminium).
- Cryolite lowers the melting point of the aluminium oxide, meaning less energy is needed (cheaper energy costs).
- Aluminium forms at the negative electrode.
- Oxygen gas forms at the positive carbon electrode and reacts with the carbon, forming carbon dioxide.
- This wears away the positive electrode, which is replaced regularly.

HT

At the cathode: $Al^{3+} + 3e^- \longrightarrow Al$

At the anode: $2O^{2-} \longrightarrow O_2 + 4e^-$

Key Point

In the exam you could be asked to suggest a hypothesis to explain given data.

Electrolysis of Aqueous Solutions

- When ionic compounds are dissolved in water to form aqueous solutions, it is slightly harder to predict the products of electrolysis.
- The water molecules break down to form hydroxide ions, OH^-, and hydrogen ions, H^+.
- At the negative electrode:
  - Hydrogen is produced if the metal is more reactive than hydrogen.
  - The metal is produced if the metal is less reactive than hydrogen.
- At the positive electrode:
  - Oxygen is produced unless the solution contains halide ions.
  - If halide ions are present, then the halogen is produced.
- In the electrolysis of sodium chloride solution:
  - Hydrogen is released at the negative electrode.
  - Chlorine gas is released at the positive electrode.

Key Point

When electrolysis is used to extract metal, the positive electrode is made of carbon.

REQUIRED PRACTICAL

Investigate what happens when aqueous solutions are electrolysed using inert electrodes.

Sample Method	Hazards and Risks
1. Set up the equipment as shown in the diagram on page 42. 2. Pass an electric current through the aqueous solution. 3. Observe the products formed at each inert electrode.	• A low voltage must be used to prevent an electric shock. • The room must be well ventilated, and the experiment must only be carried out for a short period of time, to prevent exposure to dangerous levels of chlorine gas.

Quick Test

1. Predict the products of the electrolysis of aqueous sodium bromide.
2. HT Write down the half equations for the reactions that take place at each electrode in the electrolysis of aqueous sodium bromide.

Key Words

electrolysis
electrolyte
electrode
anode
cathode
cryolite

Review Questions

Atoms, Elements, Compounds and Mixtures

1. What are the substances that are produced in a chemical reaction called? [1]
2. When magnesium metal is heated, it reacts with oxygen to form magnesium oxide.

   Write a word equation for this reaction. [1]
3. Define the term 'compound'. [1]
4. What technique could be used to separate the food colourings used to make sweets? [1]
5. When magnesium is added to hydrochloric acid a chemical reaction takes place. The products are magnesium chloride and hydrogen.

   Why is the total mass of the reactants equal to the total mass of the products in this reaction? [1]
6. Complete the following symbol equations:

   a) $H_2 + Br_2 \rightarrow$ ...... HBr [1]

   b) ...... $SO_2 + O_2 \rightarrow$ ...... SO_3 [2]

   c) $CH_4 +$ ...... $O_2 \rightarrow CO_2 +$ ...... H_2O [2]

   d) $N_2 +$ ...... $H_2 \rightarrow$ ...... NH_3 [2]

   e) ...... $K + Br_2 \rightarrow$ ...... KBr [2]

Total Marks .... / 14

Atoms and the Periodic Table

1. Complete the table to show the relative charge of the different subatomic particles.

Subatomic Particle	Relative Charge
proton	
	none
electron	

[3]

2. An atom of aluminium can be represented by: $^{27}_{13}Al$

   a) How many protons does this atom of aluminium have? [1]

   b) How many neutrons does this atom of aluminium have? [1]

   c) How many electrons does this atom of aluminium have? [1]

3. What is the typical radius of an atom?
   Tick **one** box.

   0.1nm ☐ 10nm ☐ 1mm ☐ 0.01mm ☐ [1]

4. An oxide ion can be represented by: $^{16}_{8}O^{2-}$

   a) How many protons does this oxide ion have? [1]

   b) How many neutrons does this oxide ion have? [1]

   c) How many electrons does this oxide ion have? [1]

Total Marks .... / 10

The Periodic Table

1. Copper is a transition metal.

   Give **three** properties of transition metals. [3]

2. Chlorine is a reactive non-metal.
   It has two common isotopes: chlorine-37 and chlorine-35.

   a) Which group of the periodic table does chlorine belong to? [1]

   b) What is the atomic number of chlorine? [1]

   c) Define the word 'isotope' in terms of atomic number and mass number. [1]

   d) In terms of the subatomic particles, explain the similarities and differences between an atom of chlorine-35 and an atom of chlorine-37. [3]

Total Marks .... / 9

Review Questions

States of Matter

1. State symbols are used to add extra information to equations.

   a) What does the state symbol (s) mean? [1]

   b) What does the state symbol (l) mean? [1]

Total Marks .... / 2

Ionic Compounds

1. **Table 1** shows the charge of some metal ions and non-metal ions.

Table 1

Metal Ions	Non-Metal Ions
Lithium, Li^{+}	Oxide, O^{2-}
Strontium, Sr^{2+}	Chloride, Cl^{-}
Potassium, K^{+}	Bromide, Br^{-}
Magnesium, Mg^{2+}	Sulfide, S^{2-}

   a) Define the term 'ion'. [2]

   b) In terms of electron transfer, explain why lithium ions have a 1+ charge. [2]

   c) In terms of electron transfer, explain why oxide ions have a 2– charge. [2]

   d) Use the table above to suggest the formula of:

      i) Strontium chloride. [1]

      ii) Potassium bromide. [1]

      iii) Magnesium sulfide. [1]

2. **Figure 1** shows the outer electrons in a sodium atom and in a chlorine atom.

Figure 1

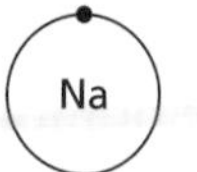

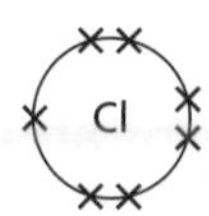

   a) i) In which group of the periodic table is sodium? [1]

      ii) In which group of the periodic table is chlorine? [1]

b) Draw a diagram to show the electronic structure of a sodium ion and a chloride ion. [2]

Show the outer shell of electrons only.

c) The compound sodium chloride is a solid at room temperature.

i) What type of bonding is present in sodium chloride? [1]

ii) Why is sodium chloride a solid at room temperature? [2]

Total Marks .... / 16

Metals

1 Lithium reacts with chlorine to produce lithium chloride:

$2Li(s) + Cl_2(g) \rightarrow 2LiCl(s)$

a) What is the state of the:

i) Lithium? **ii)** Chlorine? [2]

b) What type of bonding is present in lithium? [1]

c) Why is lithium a good electrical conductor? [2]

d) What sort of bonding is present in chlorine molecules? [1]

e) Why is chlorine a gas at room temperature? [1]

f) The reaction produces lithium chloride.

What type of bonding is present in lithium chloride? [1]

g) Lithium chloride does not conduct electricity when it is solid, but it does conduct electricity when it is molten. Explain why. [2]

2 **a)** Name the type of bonding that occurs in copper. [1]

b) Copper is used to make electrical wires because it is a very good electrical conductor.

Why is copper a good electrical conductor? [2]

Total Marks .... / 13

Review Questions

Covalent Compounds

1 **Figure 1** shows the structure of graphite.

Figure 1

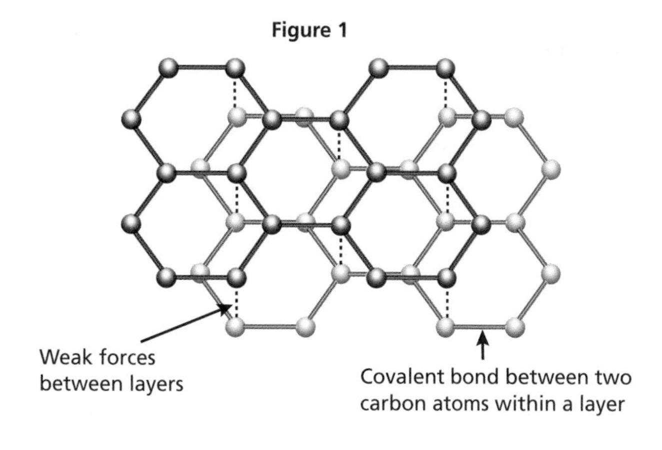

a) Graphite is a form of the element carbon.

Name **one** other form of carbon that is solid at room temperature. [1]

b) Within each layer of graphite, each carbon atom is bonded to other carbon atoms by strong bonds.

How many carbon atoms is each carbon atom joined to by strong bonds? [1]

c) Carbon in the form of graphite is the only non-metal that conducts electricity.

Explain why graphite can conduct electricity. [2]

d) Explain why graphite has a very high melting point. [2]

2 Ammonia molecules have the formula NH_3.

a) In which group of the periodic table is nitrogen? [1]

b) Ammonia has a low boiling point and is a gas at room temperature.

Explain why ammonia has a low boiling point. [2]

c) Explain why ammonia does not conduct electricity. [1]

3 **Figure 2** shows the structure of silicon dioxide.

Figure 2

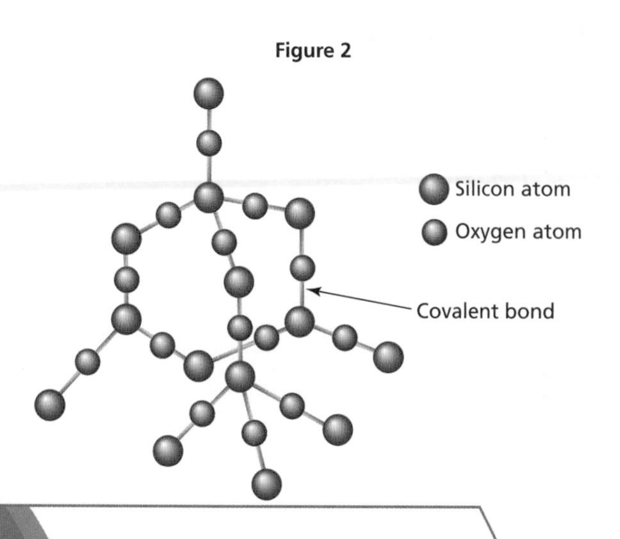

a) Silicon dioxide has the formula SiO_2.

i) How many silicon atoms is each oxygen atom bonded to? [1]

ii) How many oxygen atoms is each silicon atom bonded to? [1]

b) Silicon dioxide has a giant covalent structure.

Why does silicon dioxide have a very high melting point? [2]

4 Methane, CH_4, and butane, C_4H_{10}, are both fuels with a simple molecular structure.

a) Name the type of bonding that occurs in methane and butane molecules. [1]

b) **Figure 4** shows the electronic structure of a carbon atom and a hydrogen atom.

Figure 4

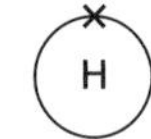

Complete **Figure 5** to show the electron arrangement in a methane molecule.

Figure 5

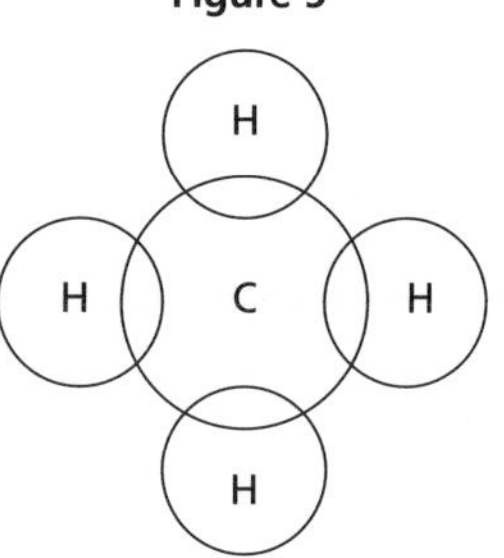

[1]

c) Which fuel – methane or butane – has the higher boiling point?
You must explain your answer. [3]

Total Marks .... / 19

Special Materials

1 Graphene was discovered in 2004.

a) What element makes up graphene? [1]

b) Describe the structure of graphene. [2]

Total Marks .... / 3

Practice Questions

Conservation of Mass

1. Sulfur dioxide is produced when sulfur is burned.
   Relative atomic masses (A_r): S = 32, O = 16

   a) Calculate the relative molecular mass of sulfur dioxide, SO_2. [2]

   b) HT Calculate the mass of 1.00 mole of sulfur dioxide, SO_2. [1]

   c) HT Calculate the mass of 0.5 moles of sulfur dioxide, SO_2. [1]

2. HT Magnesium is more reactive than copper.
   Magnesium displaces copper from a solution of copper sulfate.
   This reaction can be summed up by the balanced symbol equation below:

$$Mg + CuSO_4 \rightarrow MgSO_4 + Cu$$

   This equation can be split into two half equations.

   Complete the two half equations.

   a) $Mg \rightarrow Mg^{2+} +$ .... [1]

   b) $Cu^{2+} +$ .... $\rightarrow$ .... [2]

3. What is the relative formula mass of $Ca(NO_3)_2$?
   Relative atomic masses (A_r): Ca = 40, N = 14, O = 16
   Tick **one** box.

   164 ☐

   164g ☐

   150 ☐

   150g ☐ [1]

4. A student adds a piece of magnesium ribbon to a flask of dilute hydrochloric acid.

$$Mg(s) + 2HCl(aq) \rightarrow MgCl_2(aq) + H_2(g)$$

   Why does the mass of the reaction flask go down? [2]

Total Marks .... / 10

HT Amount of Substance

1. What unit do chemists use to measure the amount of substance?
Tick **one** box.

Grams	☐	Moles	☐
Kilograms	☐	Tonnes	☐

[1]

2. The Avogadro constant has a value of 6.02×10^{23}.

a) How many atoms are present in 7g of lithium? [1]

b) How many atoms are present in 24g of carbon? [1]

3. Calculate the number of moles in each of these substances:

a) 19g of fluorine, F_2. [2]

b) 22g of carbon dioxide, CO_2. [2]

c) 17g of hydroxide, OH^- ions. [2]

4. Complete combustion of carbon produces carbon dioxide, CO_2.

$C + O_2 \rightarrow CO_2$

1.8g of carbon was completely burned in oxygen.
Relative atomic masses (A_r): C = 12, O = 16

a) How many moles of carbon were burned? [2]

b) Calculate the mass of carbon dioxide, CO_2, produced in this reaction. [2]

5. When a hydrogen balloon explodes, the hydrogen reacts with an excess of oxygen to produce water vapour.

$2H_2 + O_2 \rightarrow 2H_2O$

1.8g of water vapour was produced in this reaction.

a) What does the term 'excess' mean? [1]

b) Calculate the amount, in moles, of water vapour produced in this reaction. [2]

Practice Questions

c) Calculate the amount, in moles, of hydrogen that reacted in this reaction. [1]

d) Calculate the mass of hydrogen that would produce 1.8g of water vapour. [2]

6 At room temperature and pressure, 1 mole of any gas takes up a volume of $24dm^3$.

a) Calculate the volume that 0.6 moles of fluorine, F_2, occupies. [2]

b) Calculate the amount, in moles, of methane, CH_4, gas present in $6dm^3$ at room temperature and pressure. [2]

7 When magnesium is heated it reacts to produce magnesium oxide.

$$2Mg(s) + O_2(g) \rightarrow 2MgO(s)$$

a) What is meant by 'conservation of mass'? [1]

b) What does the state symbol (g) mean? [1]

c) 1.2g of magnesium was heated until it had all reacted.
2.0g of magnesium oxide was produced.

Why has the mass of the solid gone up? [2]

d) Calculate the mass of oxygen that reacted with magnesium in this reaction. [1]

8 Copper hydroxide is an ionic compound.
Relative atomic masses (A_r): Cu = 63.5, O = 16, H = 1

a) Calculate the relative formula mass of copper hydroxide, $Cu(OH)_2$. [2]

b) Calculate the mass of 1.00 mole of copper hydroxide, $Cu(OH)_2$. [1]

Total Marks .... / 31

HT Titration

1 A student makes a solution of copper sulfate.
They place 0.100 mole of copper sulfate crystals in a volumetric flask.
They then add distilled water until the solution has a volume of $1.00dm^3$.

What is the concentration of this solution?
You must give the unit. [2]

2 Titration can be used to measure how much alkali is needed to neutralise an acid.

20.0cm^3 of potassium hydroxide was placed in a flask.
The potassium hydroxide has a concentration of 0.2mol/dm^3.
This required 18.0cm^3 of nitric acid solution for complete neutralisation.
The equation for the reaction can be summed up by the equation:

$$HNO_3 + KOH \rightarrow KNO_3 + H_2O$$

a) How many moles of potassium hydroxide were used in this reaction? [2]

b) How many moles of nitric acid were used in this reaction? [1]

c) What was the concentration of the nitric acid? [2]

Total Marks .... / 7

Percentage Yield and Atom Economy

1 A chemist expected a reaction to produce 12.0g of product.
However, after carrying out the reaction only 9.5g of product was actually produced.

a) Why might the actual yield of a reaction be less than the theoretical yield of the product?
Tick **one** box.

The reaction is reversible and does not go to completion. ☐

One reactant is in excess. ☐

One reactant is limiting the reaction. ☐

A reaction has an atom economy of less than 100%. ☐ [1]

b) Calculate the percentage yield of this reaction.
Give your answer to the nearest whole number. [2]

Total Marks .... / 3

Practice Questions

Reactivity of Metals

1 Calcium can be burned to produce calcium oxide:

$$2Ca + O_2 \rightarrow 2CaO$$

a) Write a word equation for this reaction. [1]

b) HT The balanced symbol equation can be broken into two ionic equations:

$$2Ca \rightarrow 2Ca^{2+} + 4e^-$$
$$O_2 + 4e^- \rightarrow 2O^{2-}$$

In terms of oxidation and reduction, explain what happens to the calcium and the oxygen in the reaction. [4]

Total Marks .... / 5

The pH Scale and Salts

1 Which of these values shows the pH of a strong alkali?
Tick **one** box.

7 ☐ 14 ☐ 8  1 ☐ [1]

2 Which of these ions is found in excess in acidic solutions?
Tick **one** box.

H^+ ☐ H^- ☐ OH^+  OH^- ☐ [1]

3 Sulfuric acid is a strong acid.

a) A solution of sulfuric acid has a pH of 1.

What does the pH scale measure? [2]

b) Complete the equation to show the neutralisation reaction between sulfuric acid and potassium hydroxide.

sulfuric acid + potassium hydroxide → .... + ... [2]

Total Marks .... / 6

Electrolysis

1. Copper chloride is an ionic compound that can be separated by electrolysis.

   a) Name the element formed during the electrolysis of molten copper chloride:

      i) At the positive electrode. [1]

      ii) At the negative electrode. [1]

   b) HT Complete the half equations to show the reactions that take place at each electrode.

      i) At the anode: $2Cl^- \rightarrow$ .... $+ 2e^-$ [1]

      ii) At the cathode: $Cu^{2+} +$ .... $\rightarrow Cu$ [1]

2. A student carries out an experiment to find out what happens when an aqueous solution of sodium chloride is electrolysed.

   a) Identify the two positive ions present in an aqueous solution of sodium chloride. [2]

   b) Name the substance produced at the negative electrode. [2]
   You must give a reason for your answer.

   c) Name the substance produced at the positive electrode. [2]
   You must give a reason for your answer.

3. During the electrolysis of molten lead chloride, lead and chlorine are produced.

   At the anode: $2Cl^- \rightarrow Cl_2 + 2e^-$
   At the cathode: $Pb^{2+} + 2e^- \rightarrow Pb$

   a) HT During the electrolysis of lead chloride, oxidation and reduction reactions take place.

      i) In terms of oxidation and reduction, describe what happens to the lead ions. [2]

      ii) In terms of oxidation and reduction, describe what happens to the chloride ions. [2]

   b) Why does the lead chloride have to be molten? [1]

Total Marks .... / 15

Exothermic and Endothermic Reactions

You must be able to:

- Recall that exothermic reactions transfer energy to the surroundings and result in an increase in temperature
- Recall that endothermic reactions take in energy from the surroundings and result in a decrease in temperature
- Give some examples of exothermic and endothermic reactions.

Energy Transfers

- When chemical reactions occur, energy is transferred from the chemicals to or from the surroundings. Therefore, many reactions are accompanied by a temperature change.
- **Exothermic reactions** are accompanied by a temperature rise.
- They transfer heat energy from the chemicals to the surroundings, i.e. they give out heat energy.
- Exothermic reactions are used in products like self-heating cans (for coffee) and hand warmers.
- **Endothermic reactions** are accompanied by a fall in temperature.
- Heat energy is transferred from the surroundings to the chemicals, i.e. they take in heat energy.
- Some sports injury packs use endothermic reactions.
- If a **reversible reaction** is exothermic in one direction, then it is endothermic in the opposite direction.
- The same amount of energy is transferred in each case.

REQUIRED PRACTICAL

Investigate the variables that affect temperature changes in reacting solutions.

Sample Method	**Considerations, Mistakes and Errors**
1. Set up the equipment as shown. 2. Take the temperature of the acid. 3. Add the metal powder and stir. 4. Record the highest temperature the reaction mixture reaches. 5. Calculate the temperature change for the reaction. 6. Repeat the experiment using a different metal.	• There should be a correlation between the reactivity of the metal and the temperature change, i.e. the more reactive the metal, the greater the temperature change. • When a measurement is made there is always some uncertainty about the results obtained. For example, if the experiment is repeated three times and temperature changes of 3°C, 4°C and 5°C are recorded: - the range of results is from 3°C to 5°C - the mean (average) = $\frac{(3 + 4 + 5)}{3}$ = 4°C
Variables	**Hazards and Risks**
• The independent variable is the metal used. • The dependent variable is the temperature change. • The control variables are the type, concentration and volume of acid.	• There is a low risk of a corrosive acid damaging the experimenter's eye, so eye protection must be used.

Energy Level Diagrams

- In chemical reactions, atoms are rearranged as old bonds are broken and new bonds are formed.

Key Point

Exothermic reactions are accompanied by a temperature rise.

Endothermic reactions are accompanied by a fall in temperature.

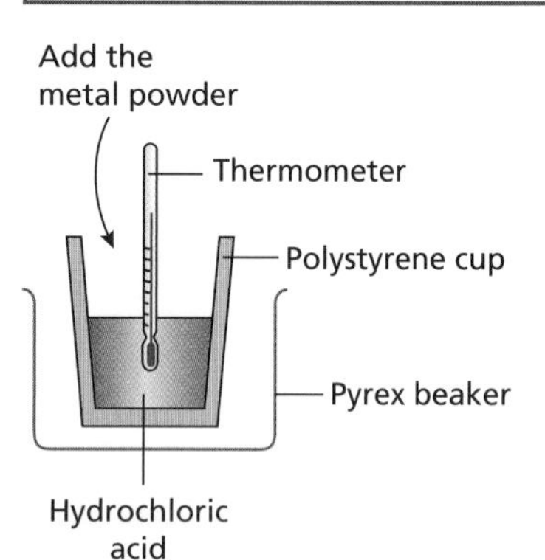

Key Point

An accurate measurement is one that is close to the true value. If the experiment was repeated the results could be analysed to see if they are precise. The results are precise if they are all close together.

Measurements are repeatable if the experimenter repeats the experiment and gets similar results. Results are reproducible if similar results are obtained by different experimenters using different equipment.

- For bonds to be broken, reacting particles must collide with sufficient energy.
- The minimum amount of energy that the particles must have for a reaction to take place is called the **activation energy**.
- The energy changes in a chemical reaction can be shown using an **energy level diagram** or **reaction profile**.
- In an exothermic reaction:
  - energy is given out to the surroundings
  - the products have less energy than the reactants.
- In an endothermic reaction:
  - energy is being taken in from the surroundings
  - the products have more energy than the reactants.

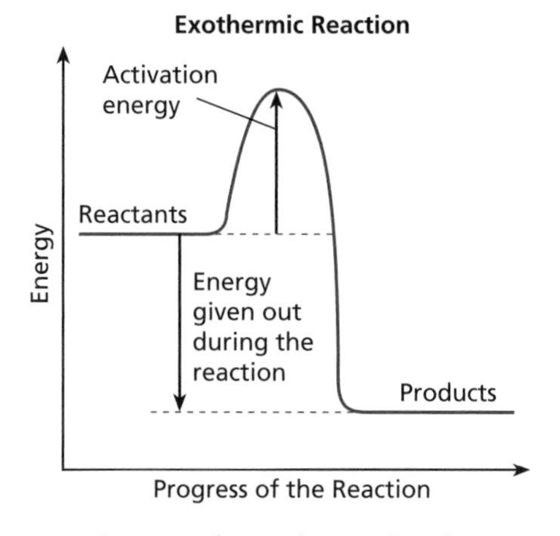

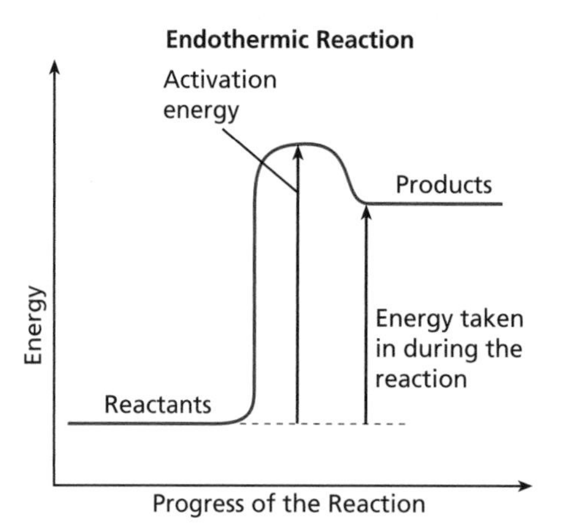

- **Catalysts** reduce the activation energy needed for a reaction. This makes the reaction go faster.

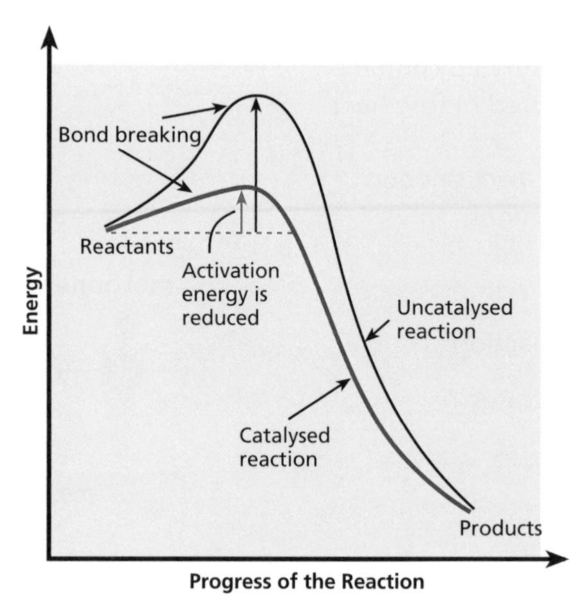

Quick Test

1. What is an endothermic reaction?
2. A white powder dissolves in a solution. The temperature falls by 2°C. State and explain the type of reaction that has taken place.
3. Sketch the energy level diagram for an:
   a) exothermic reaction
   b) endothermic reaction.

Key Words

exothermic reaction
endothermic reaction
reversible reaction
activation energy
energy level diagram
reaction profile
catalyst

Fuel Cells

You must be able to:

- HT Recall energy is released when new bonds are made
- HT Calculate the energy transferred in reactions and use it to deduce whether a reaction is exothermic or endothermic
- Understand how simple cells can be made
- Explain how hydrogen fuel cells work and why they are so useful
- HT Write half equations for the electrode reactions in the hydrogen fuel cell.

HT Energy Change of Reactions

- In a chemical reaction, new substances are produced:
  - The bonds in the reactants are broken.
  - New bonds are made to form the products.
- Breaking a chemical bond requires energy – it is an **endothermic** process.
- When a new chemical bond is formed, energy is given out – it is an **exothermic** process.
- If more energy is required to break bonds than is released when bonds are formed, the reaction must be endothermic.
- If more energy is released when bonds are formed than is needed to break bonds, the reaction must be exothermic.

HT Key Point

Endothermic:
energy required to break old bonds > energy released when new bonds are formed

Exothermic:
energy required to break old bonds < energy released when new bonds are formed

Measuring Energy Changes

- The amount of energy produced in a chemical reaction in solution can be measured by mixing the reactants in an insulated container.
- This enables the temperature change to be measured before heat is lost to the surroundings.
- This method would be suitable for neutralisation reactions and reactions involving solids, e.g. metal and acid reactions.

Energy Calculations

Calculate the energy transferred in the following reaction:

methane + oxygen → carbon dioxide + water

$$CH_4(g) + 2O_2(g) \rightarrow CO_2(g) + 2H_2O(g)$$

The bond energies needed for this are:

C–H is 412kJ/mol, O=O is 496kJ/mol
C=O is 805kJ/mol, H–O is 463kJ/mol

Energy used to break bonds is:
4 C–H = 4 × 412 = 1648kJ
2 O=O = 2 × 496 = 992kJ
Total = 1648kJ + 992kJ = 2640kJ

Energy given out by making bonds:
2 C=O = 2 × 805 = 1610kJ
4 H–O = 4 × 463 = 1852kJ
Total = 1610kJ + 1852kJ = 3462kJ

energy change =
energy used to break bonds – energy given out by making bonds

= 2640kJ – 3462kJ = –822kJ

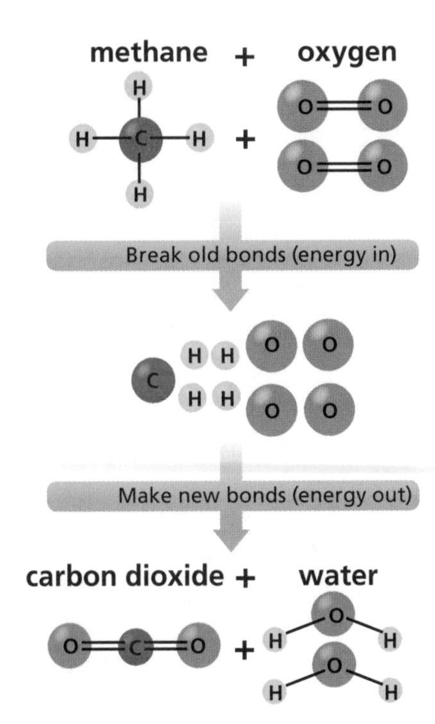

Cells and Batteries

- The chemicals in **cells** react together to release electricity.
- A simple cell can be made by placing two different metals into a beaker containing an electrolyte.
- Batteries are made when two or more cells are connected together.
- Cells can be joined in series to produce a higher voltage, e.g. if two 3.0V cells are joined in series, the total voltage produced by the battery is 6.0V.

A Simple Cell

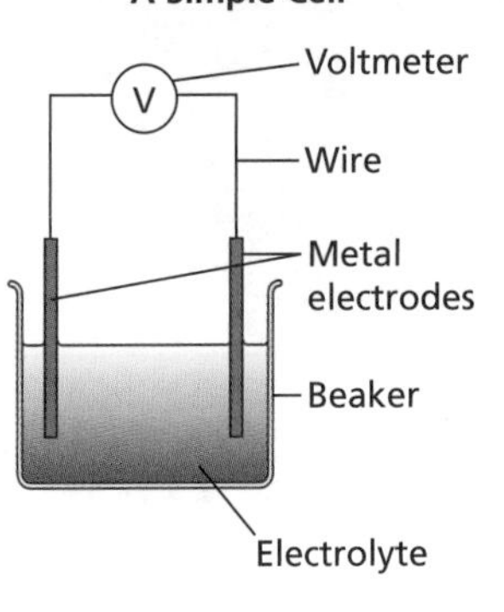

Fuel Cells

- **Fuel cells** are a very efficient way of producing electrical energy.
- Most fuel cells use hydrogen, but other fuels can be used.
- In a hydrogen fuel cell:
  - Hydrogen and oxygen are supplied to the fuel cell.
  - The fuel is oxidised to produce a potential difference (voltage).
  - Overall, the hydrogen is oxidised to form water.

HT Hydrogen is added at the anode.

HT The hydrogen molecules lose electrons to form hydrogen ions.

$$2H_2 \longrightarrow 4H^+ + 4e^-$$

HT Oxygen is added at the cathode.

HT The hydrogen ions formed at the anode travel through the electrolyte to the cathode.

HT At the cathode, the hydrogen ions react with oxygen molecules to form water.

$$O_2 + 4H^+ + 4e^- \longrightarrow 2H_2O$$

HT The overall equation for the reaction in the hydrogen fuel cell is found by adding the two half equations together:

$$2H_2 + O_2 \longrightarrow 2H_2O$$

- Traditional methods of producing electricity involve many stages and are less efficient and more polluting than fuel cells.
- Fuel cells have lots of advantages:
  - Hydrogen fuel cells produce water, which is non-polluting.
  - They are lightweight and small.
  - They have no moving parts so are very unlikely to break down.

Key Point

Rechargeable cells and batteries can be recharged because the chemical reactions that take place within them are reversed when an external electrical current is supplied.

Key Point

Fuel cells are a very efficient way of producing electrical energy.

The fuel is oxidised electrochemically to produce a potential difference or voltage.

Quick Test

1. HT A chemical reaction gives out more energy when bonds are made than it takes in to break bonds. What sort of reaction is it?
2. List **four** advantages of hydrogen fuel cells over traditional ways of producing electricity.
3. HT Give the equation for the chemical reaction which takes place at:
   a) the anode of a hydrogen fuel cell
   b) the cathode of a hydrogen fuel cell.

Key Words

HT endothermic
HT exothermic
cell
fuel cell

Rate of Reaction

You must be able to:

- Describe how the rate of a chemical reaction can be found
- Use collision theory to explain how factors affect the rate of reactions
- HT Calculate the rate of a reaction from graphs.

Calculating the Rate of Reaction

rate of reaction = $\frac{\text{amount of reactant used OR product formed}}{\text{time}}$

- The rate of reaction can be found in different ways.
- **Measuring the amount of reactants used:**
  - If one of the products is a gas, measure the mass in grams (g) of the reaction mixture before and after the reaction takes place and the time it takes for the reaction to happen.
  - The mass of the mixture will decrease.
  - The units for the rate of reaction may then be given as g/s.

  HT The amount of a reactant can also be measured in moles (mol).
  HT As the reaction takes place the reactant is used up, so the amount of reactant remaining decreases.
  HT The concentration of the reactant is calculated as the amount (mol) divided by the volume of the reaction mixture (dm^3). It is measured in units of mol/dm^3.

- **Measuring the amount of products formed:**
  - If one of the products is a gas, measure the total volume of gas produced in cubic centimetres (cm^3) with a gas syringe and the time it takes for the reaction to happen.
  - The units for the rate of reaction may then be given as cm^3/s.
- **Measuring the time it takes for a reaction mixture to become opaque or change colour:**
  - Time how long it takes for the mixture to change colour.
  - Rate of reaction $\approx \frac{1}{\text{time taken for solution to change colour}}$.

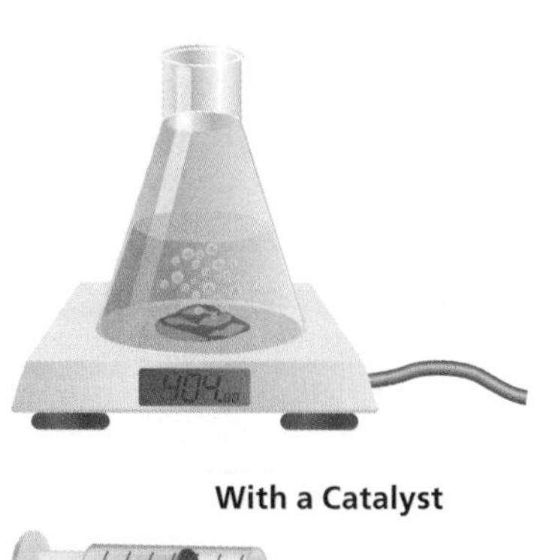

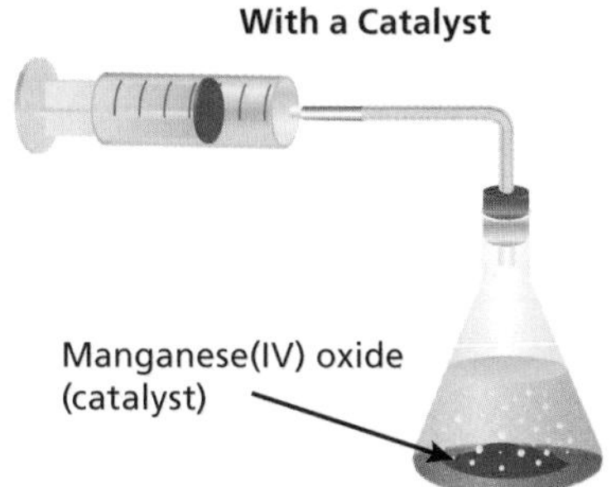

With a Catalyst

Collision Theory

- Chemical reactions only occur when reacting particles collide with each other with sufficient energy.
- The minimum amount of energy required to cause a reaction is called the activation energy.
- There are four important factors that affect the rate of reaction: temperature, concentration, surface area and catalysts (see page 62).
- **Temperature**:
  - In a hot reaction mixture the particles move more quickly – they collide more often and with greater energy, so more collisions are successful.

- **Concentration**:
  - At higher concentrations, the particles are crowded closer together – they collide more often, so there are more successful collisions.
  - Increasing the pressure of reacting gases also increases the frequency of collisions.

REQUIRED PRACTICAL **Investigate how changes in concentration affect the rates of reactions by methods involving the production of gas or a colour change.**	
This investigation uses the reaction between sodium thiosulfate and hydrochloric acid. **Sample Method** **1.** Set up the equipment as shown. **2.** Add the hydrochloric acid to the flask and swirl to mix the reactants. **3.** Start the timer. **4.** Watch the cross through the flask. **5.** When the cross is no longer visible stop the timer. **6.** Repeat the experiment using hydrochloric acid of a different concentration.	**Considerations, Mistakes and Errors** • There should be a correlation between the concentration of the acid and the time taken for the cross to 'disappear'. • The higher the concentration of the acid, the faster the rate of reaction, and the shorter the time for the cross to 'disappear'.
Variables • The independent variable is the concentration of the acid. • The dependent variable is the time it takes for the cross to 'disappear'. • The control variables are the volume of acid and the concentration and volume of sodium thiosulfate.	**Hazards and Risks** • Corrosive acid can damage eyes, so eye protection must be used. • Sulfur dioxide gas can trigger an asthma attack, so the temperature must always be kept below 50°C.

- **Surface area**:
  - Small pieces of a solid reactant have a large surface area in relation to their volume.
  - More particles are exposed and available for collisions, so there are more collisions and a faster reaction.

Key Point

Smaller pieces have a higher surface area to volume ratio than larger pieces.

Plotting Reaction Rates

- Graphs can be plotted to show the progress of a chemical reaction.
- There are three key things to remember:
  - The steeper the line, the faster the reaction.
  - When one of the reactants is used up, the reaction stops (the line becomes horizontal).
  - The same amount of product is formed from the same amount of reactants, regardless of rate.

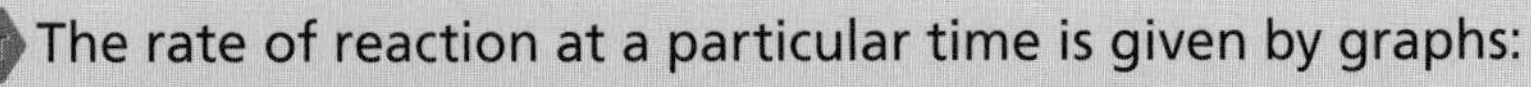

HT The rate of reaction at a particular time is given by graphs:
- Draw a **tangent** to the curve at that time.
- Find the **gradient** of the tangent.
- The gradient is equal to the rate of reaction at that time.

$$\text{gradient} = \frac{\text{difference in the amount of product formed / reactant used}}{\text{time}}$$

The graph shows that reaction A is faster than reaction B.

Quick Test

1. Why does increasing temperature increase the rate of reaction?
2. What is the name given to the minimum amount of energy that reacting particles must have to react?
3. What is a catalyst?

Key Words

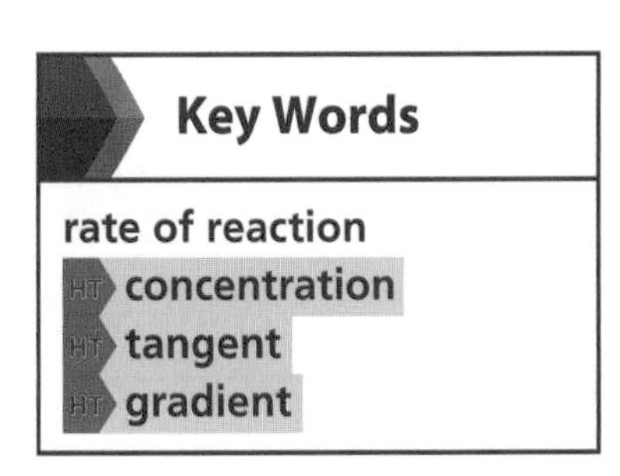

rate of reaction
HT **concentration**
HT **tangent**
HT **gradient**

Reversible Reactions

You must be able to:

- Explain how and why catalysts can affect the rate of reaction
- Explain what a reversible reaction is
- Define the term 'equilibrium'
- HT Predict the effect of changing the conditions on a system at equilibrium.

Catalysts

- A **catalyst** is a substance that increases the rate of a chemical reaction without being used up in the process.
- Catalysts are not included in the chemical equation for the reaction.
- A catalyst:
  - reduces the amount of energy needed for a successful collision
  - makes more collisions successful
  - speeds up the reaction
  - provides a surface for the molecules to attach to, which increases their chances of bumping into each other.
- Enzymes act as catalysts in biological systems.
- Different reactions need different catalysts, e.g.
  - the cracking of hydrocarbons uses broken pottery
  - the manufacture of ammonia uses iron.
- Increasing the rates of chemical reactions is important in industry, because it helps to reduce costs.

Catalysts Used in Industrial Reactions

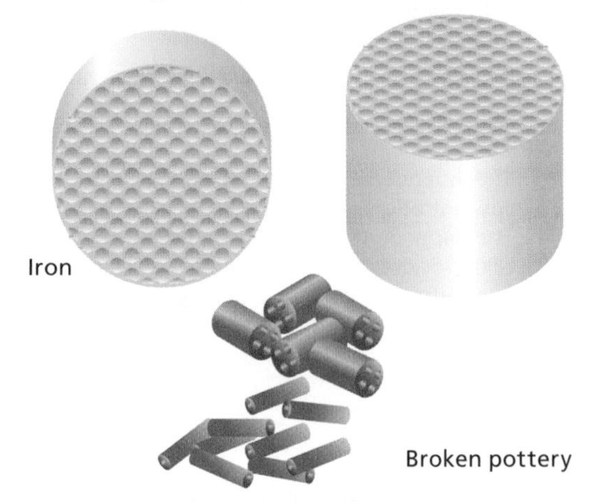

Reversible Reactions

- Some chemical reactions are reversible, they can go forwards or backwards.
- In a **reversible reaction**, the products can react to produce the original reactants.
- These reactions are represented as:

$$A(g) + B(g) \rightleftharpoons C(g) + D(g)$$

- This means that:
  - A and B can react to produce C and D.
  - C and D can react to produce A and B.
- For example:
  - Solid ammonium chloride decomposes when heated to produce ammonia and hydrogen chloride gas (both colourless).
  - Ammonia reacts with hydrogen chloride gas to produce clouds of white ammonium chloride.

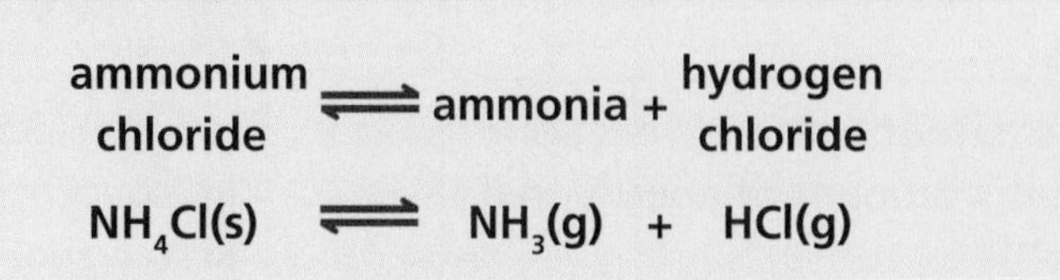

ammonium chloride ⇌ ammonia + hydrogen chloride

$$NH_4Cl(s) \rightleftharpoons NH_3(g) + HCl(g)$$

Key Point

Some chemical reactions are reversible, they can go forwards or backwards.

- The direction of reversible reactions can be changed by changing the conditions.

Closed Systems

- In a **closed system**, no reactants are added and no products are removed.
- When a reversible reaction occurs in a closed system, an **equilibrium** is achieved when the rate of the forward reaction is equal to the rate of the backward reaction.
- The relative amounts of all the reacting substances at equilibrium depend on the conditions of the reaction.

HT Changing Reaction Conditions

- **Le Chatelier's Principle** states that if a system in equilibrium is subjected to a change in conditions, then the system shifts to resists the change.
- In an exothermic reaction:
  - If the temperature is raised, the yield decreases.
  - If the temperature is lowered, the yield increases.
- In an endothermic reaction:
  - If the temperature is raised, the yield increases.
  - If the temperature is lowered, the yield decreases.
- In reactions involving gases:
  - An increase in pressure favours the reaction that produces the least number of gas molecules.
  - A decrease in pressure favours the reaction that produces the greater number of gas molecules.
- If the concentration of one of the reactants or products is changed:
  - the system is no longer in equilibrium
  - the system adjusts until it can reach equilibrium once more.
- If the concentration of one of the reactants is increased, the position of equilibrium shifts so that more products are formed until equilibrium is reached again.
- In contrast, if the concentration of one of the reactants is decreased, the position of equilibrium shifts so that more reactants are formed until equilibrium is reached again.
- These factors, together with reaction rates, determine the optimum conditions in industrial processes, e.g. the Haber process (see pages 96–97).

Key Point

When a reversible reaction occurs in a closed system, an equilibrium is achieved when the rate of the forward reaction is exactly the same rate as the backward reaction.

Quick Test

1. What is a reversible reaction?
2. When is a system in equilibrium?
3. HT How can the effect of changing the conditions in a system that is in equilibrium be predicted?

Key Words

catalyst
reversible reaction
closed system
equilibrium
HT Le Chatelier's Principle

Alkanes

You must be able to:

- Describe how crude oil is formed
- Recall the general formula for alkanes
- Understand how the fractional distillation of crude oil works
- Recall how the properties of hydrocarbons are linked to their size.

Crude Oil and Hydrocarbons

- **Crude oil** is:
  - formed over millions of years from the fossilised remains of plankton
  - found in porous rocks in the Earth's crust
  - a finite (non-renewable) resource that is used to produce fuels and other chemicals.
- Most of the compounds in crude oil are **hydrocarbons** (molecules made of only carbon and hydrogen atoms).
- Hydrocarbon molecules vary in size, which affects their properties and how they can be used as fuels.
- The larger the hydrocarbon:
  - the more **viscous** it is (i.e. the less easily it flows)
  - the higher its boiling point
  - the less volatile it is
  - the less easily it ignites.

Key Point

Fuels are substances that can be burned to release energy.

Fractional Distillation

- Crude oil can be separated into different fractions (parts) by **fractional distillation**.
- Each fraction contains hydrocarbon molecules with a similar number of carbon atoms.
- Most of the hydrocarbons obtained are **alkanes** (see below).
- First, the crude oil is heated until it evaporates.
- The vapour moves up the fractionating column.
- The top of the column is much colder than the bottom.
- Shorter hydrocarbon molecules can reach the top of the fractionating column before they condense and are collected.
- Longer hydrocarbon molecules condense at higher temperatures and are collected lower down the column.

Fractional Distillation

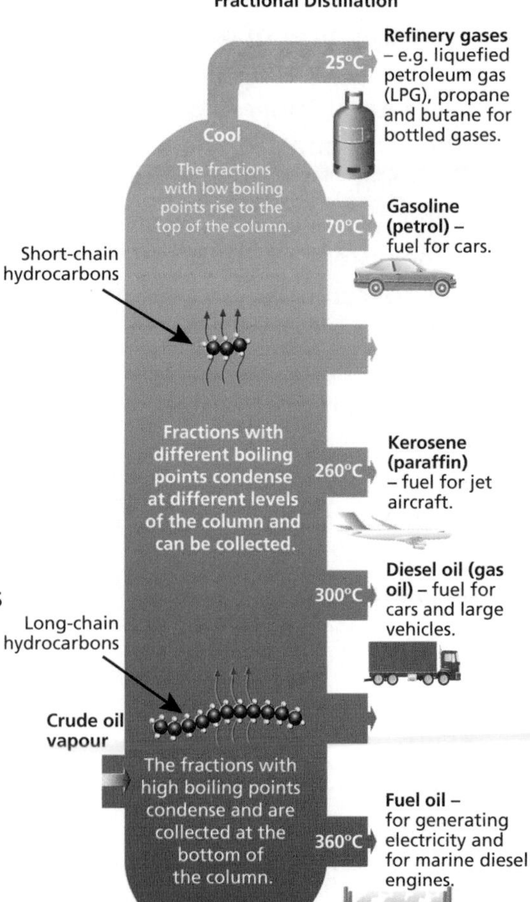

Alkanes

- Carbon atoms are linked to four other atoms by single bonds.
- Alkanes only contain single bonds and are described as **saturated** hydrocarbons.
- Alkanes are fairly unreactive, but they burn well.
- The general formula for alkanes is:

LEARN

$$C_nH_{2n+2}$$

- Alkanes can be drawn with a single line between atoms, which represents a single covalent bond:

−H	−C−	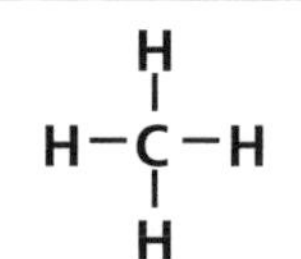	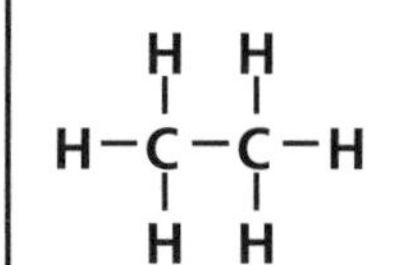	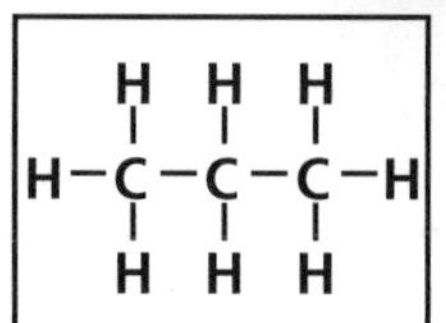
Hydrogen atoms can make 1 bond each.	Carbon atoms can make 4 bonds each.	The simplest alkane, **methane**, **CH_4**, is made up of 4 hydrogen atoms and 1 carbon atom.	**Ethane, C_2H_6** A molecule made up of 2 carbon atoms and 6 hydrogen atoms.	**Propane, C_3H_8** A molecule made up of 3 carbon atoms and 8 hydrogen atoms.

- The shorter-chain alkanes release energy more quickly by burning, so there is greater demand for them as fuels.

Burning Fuels

- Most fuels are compounds of carbon and hydrogen. Many also contain sulfur.
- During the **combustion** (burning) of hydrocarbon fuels:
  - both carbon and hydrogen are oxidised
  - energy is released
  - waste products are produced, which are released into the atmosphere.
- If combustion is not complete, then carbon monoxide, unburnt fuels and solid particles containing soot (carbon) may be released.
- Carbon monoxide is a colourless, odourless and toxic gas.
- Solid particles in the air, called particulates, can cause global dimming by reducing the amount of sunlight reaching the Earth's surface and cause damage to people's lungs.
- Due to the high temperature reached when fuels burn, nitrogen in the air can react with oxygen to form nitrogen oxides.
- These gases can cause respiratory problems in people and react with rain water (in the same way as sulfur dioxide) to form acid rain, which can damage plants and buildings.
- Sulfur can be removed from fuels before burning (in motor vehicles) and removed from the waste gases after combustion (in power stations).

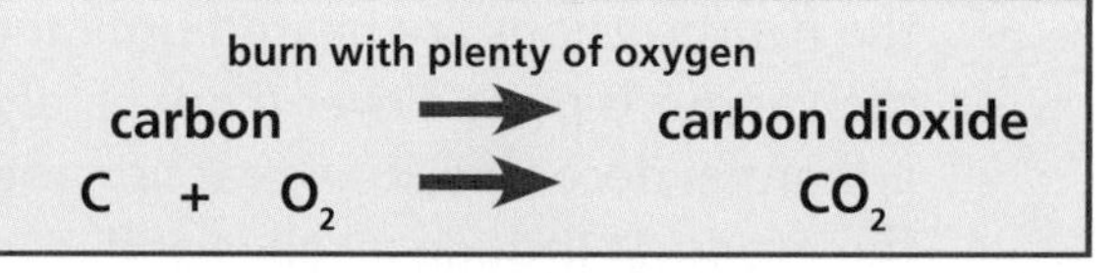

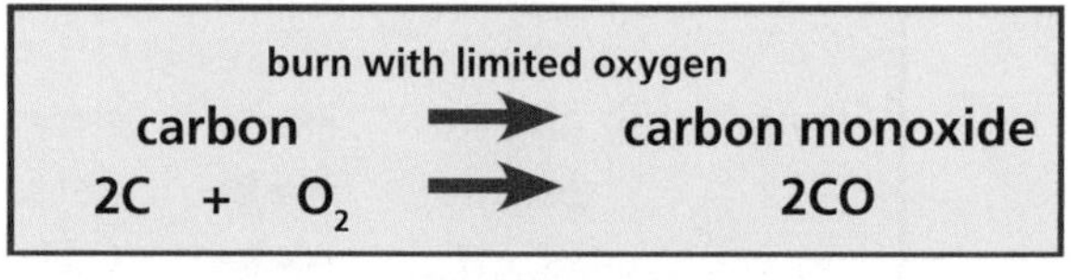

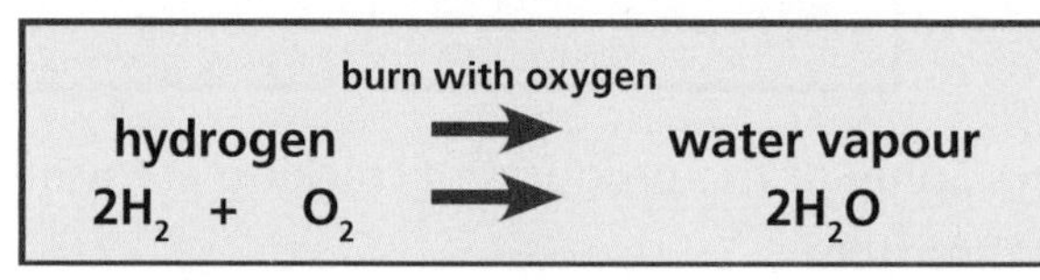

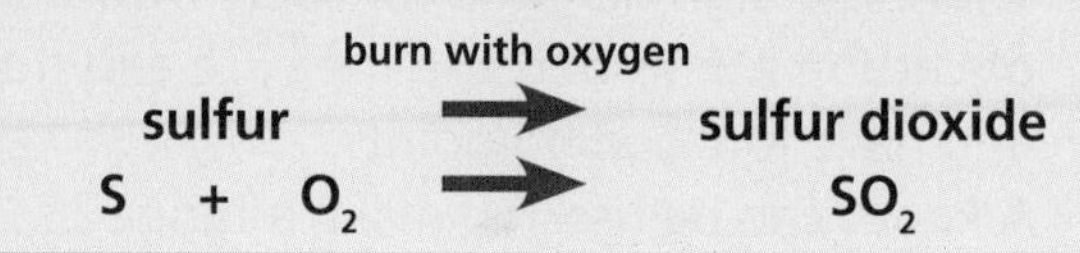

Quick Test

1. What is a hydrocarbon?
2. How many bonds does a carbon atom form?
3. What is the chemical formula for propane?
4. What is the general formula of an alkane?
5. What is the balanced symbol equation for the complete combustion of methane?

Key Words

crude oil
hydrocarbon
viscous
fractional distillation
alkanes
saturated
combustion

Alkenes

You must be able to:

- Recall the general formula for alkenes
- Describe the cracking of alkanes
- Describe the combustion of alkenes
- Understand the addition reactions of alkenes with water, hydrogen and halogens.

Cracking Hydrocarbons

- Longer-chain hydrocarbons can be broken down into shorter, more useful hydrocarbons. This process is called **cracking**.
- During cracking:
  - the hydrocarbons are heated until they vaporise
  - the vapour is passed over a hot catalyst
  - a thermal decomposition reaction then takes place
  - the products include alkanes and **alkenes**.

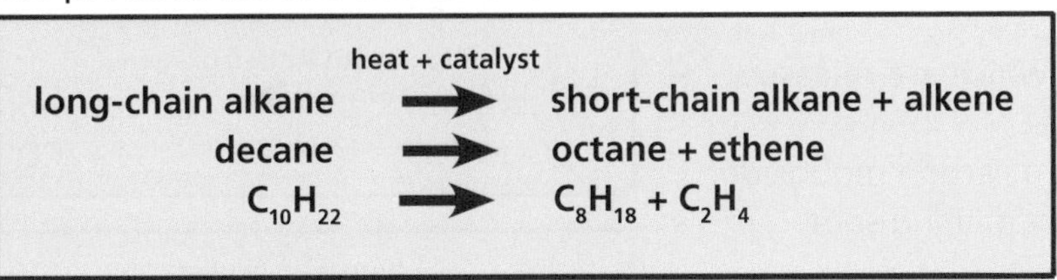

- Some of the products of cracking are useful as fuels.
- There is a high demand for fuels with small chains of carbon atoms because they are easy to ignite and have low boiling points.
- Alkenes can be used to make a range of new compounds, including polymers and industrial alcohol.

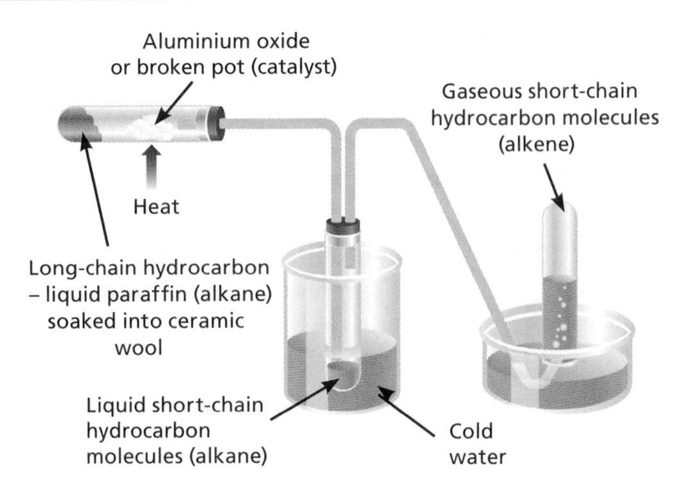

Alkenes

- As well as forming single bonds with other atoms, carbon atoms can form double bonds.
- This means that:
  - **not all the carbon atoms have to be linked to four other atoms**
  - **a double carbon–carbon (C=C) bond can be present instead.**
- Hydrocarbons that have double bonds are described as **unsaturated**.
- Alkenes have at least one double bond, so are unsaturated.
- The general formula for alkenes is:

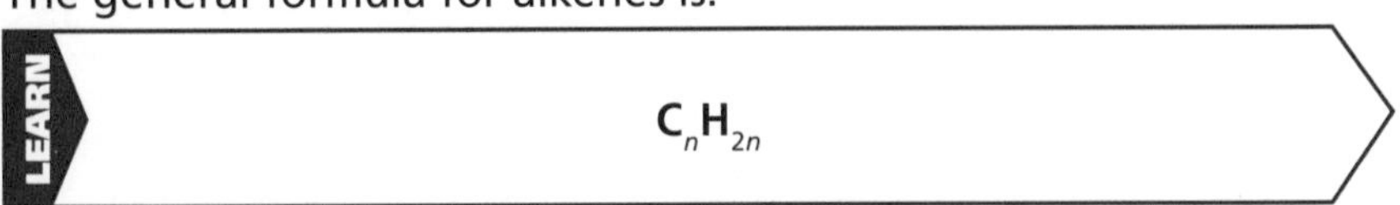

- The simplest alkene is ethene, C_2H_4.
- Ethene is made up of four hydrogen atoms and two carbon atoms and contains one double carbon–carbon (C=C) bond.

- Alkenes can be represented using displayed formulae:

Alkene	Ethene, C_2H_4	Propene, C_3H_6	Butene, C_4H_8
Displayed Formula	H H \ / C=C / \ H H	H H \ | C=C−C−H / | | H H H	H H H \ | | C=C−C−C−H / | | | H H H H

Reactions of Alkenes

- Alkenes are more reactive than alkanes due to the C=C bonds.
- Alkenes react with oxygen in combustion reactions.
- They tend to burn with smokier flames than alkanes due to incomplete combustion.
- Hydrogen can be added to alkenes to produce alkanes.
- A nickel catalyst is used.

propene + hydrogen → propane
$C_3H_6 + H_2 \rightarrow C_3H_8$

- This is an **addition reaction**.
- Ethanol (an alcohol) can be produced by reacting ethene with steam in the presence of a catalyst, phosphoric acid.

ethene + steam → ethanol
$C_2H_4 + H_2O \rightarrow C_2H_5OH$

- This is an addition reaction.

Key Point

Hydrocarbons that have double bonds are unsaturated and have the capacity to make further bonds. Alkenes are unsaturated hydrocarbons.

Bromine Water

- Alkenes are more reactive than alkanes.
- They react when shaken with bromine water, turning it from orange to colourless.
- This can be used to differentiate between alkanes and alkenes.

ethene (colourless) + bromine water (orange brown) → colourless solution
ethane (colourless) + bromine water (orange brown) → orange brown solution

- Ethene reacts with bromine to form dibromoethane in an addition reaction:

$C_2H_4 + Br_2 \rightarrow CH_2BrCH_2Br$

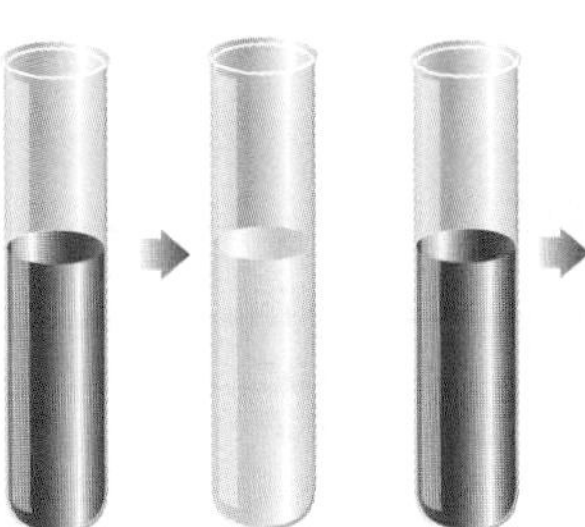

Key Point

Other halogens react with alkenes in a similar way to bromine.

Quick Test

1. Why are alkenes more reactive than alkanes?
2. Complete the equation for the cracking of octane: $C_8H_{18} \rightarrow C_6H_{14} +$ ....
3. Pentene has five carbon atoms. Give the formula of pentene.
4. Why do alkenes burn with smoky flames?
5. Explain how a sample of butene can be told apart from a sample of butane.

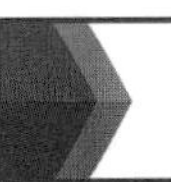

Key Words

cracking
alkenes
unsaturated
addition reaction

Organic Compounds

You must be able to:

- Recall the conditions needed for the fermentation of sugar
- Recall that all alcohols have the same functional group and some of the reactions of alcohols
- Recall that all carboxylic acids have the same functional group and some of the reactions of carboxylic acids
- Describe how esters are made
- Describe the structures of amino acids, DNA, starch and cellulose.

Fermentation

- Aqueous solutions of ethanol can be produced by the **fermentation** of sugar, which is a renewable resource.
- During fermentation:

sugar ⟶ ethanol + carbon dioxide

- Temperatures of 25°C to 50°C work best. If the temperature is:
  - too low – the yeast becomes inactive and the rate of reaction slows
  - too high – the yeast is denatured and stops working.

Key Point

The functional group is the part of a molecule that gives it its characteristic properties.

Key Point

The members of a homologous group have the same functional group, so they all react in a similar way.

Alcohols

- Alcohols are carbon-based molecules that contain the functional group **hydroxyl**, –OH.
- Methanol, ethanol, propanol and butanol are the first four members of the **homologous series** of alcohols.

Alcohol	Structural Formula	Formula
Methanol	H–C(H)(H)–O–H	**CH_3OH**
Ethanol	H–C(H)(H)–C(H)(H)–O–H	**CH_3CH_2OH**

- Alcohols:
  - dissolve in water to form neutral solutions
  - react with sodium to produce hydrogen
  - burn in air to produce carbon dioxide and water
  - are used as fuels and solvents.
- Alcoholic drinks contain ethanol.
- Ethanol can be oxidised to ethanoic acid by chemical oxidising agents or by the action of bacteria from the air.
- Ethanoic acid is the main acid in vinegar.

Key Point

Aqueous solutions of weak acids have a higher pH than aqueous solutions of strong acids with the same concentration.

Carboxylic Acids

- Carboxylic acids are organic compounds that contain the functional group **carboxyl**, –COOH.

- Carboxylic acids:
  - dissolve in water to form acidic solutions
  - react with carbonates (e.g. sodium carbonate) to produce carbon dioxide
  - react with alcohols (in the presence of an acid catalyst) to form esters
  - do not ionise (dissociate) fully in water, so they are called weak acids.

Esters

- Alcohols and carboxylic acids react together to form esters.
- Esters contain the functional group –COO.
- When ethanol and ethanoic acid react together, the ester formed is ethyl ethanoate.
- Esters are volatile compounds, i.e. they have a low boiling point.
- They have distinctive smells and are used in perfumes and as flavourings in food.

HT Amino Acids

- **Amino acids** contain two different functional groups:
  - the amine group, NH_2
  - the carboxyl group, COOH.
- Glycine, NH_2CH_2COOH, is an amino acid.
- Different amino acids join together to form polymers (see pages 70–71) called proteins.

$$H_2N-\underset{\displaystyle H}{\overset{\displaystyle H}{C}}-COOH$$

DNA (Deoxyribonucleic Acid)

- **DNA** is a very large molecule.
- It is essential for life – it stores and transmits the instructions for the development of living organisms and some viruses.
- DNA is made from two polymer chains constructed from four different nucleotides: cytosine (C), guanine (G), adenine (A) and thymine (T).
- The two polymer chains form a double helix (spiral).

Starch and Cellulose

- Starch and cellulose are polymers of sugars.
- They are made by plants and are important for life.
- Sugar, starch and **cellulose** are all carbohydrates.

Quick Test

1. What is the ideal temperature range for the fermentation of glucose?
2. Give the formula of the fourth member of the homologous series of alcohols, butanol.
3. Give the formula of the fourth member of the homologous series of carboxylic acids, butanoic acid.
4. Name the products of the complete combustion of ethanol.

Methanoic Acid

HCOOH

Ethanoic Acid

CH_3COOH

Propanoic Acid

C_2H_5COOH

Ethyl Ethanoate ($CH_3COOC_2H_5$)

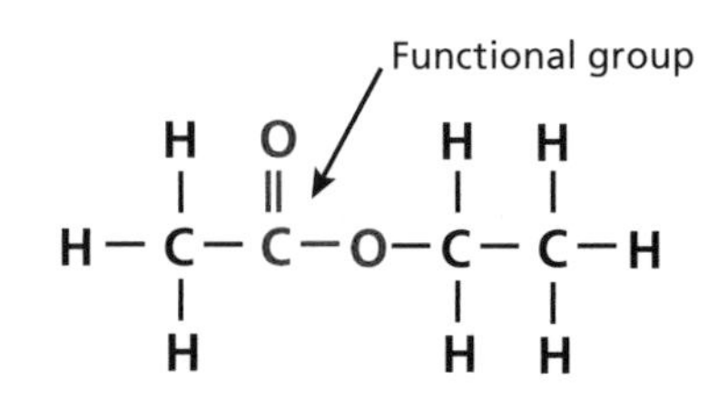

Key Point

Glucose molecules join together to form starch molecules.

Key Words

fermentation
hydroxyl
homologous series
carboxyl
HT amino acid
DNA
cellulose

Polymerisation

You must be able to:

- Describe how monomers join together in addition polymerisation reactions
- Understand how polymers are represented
- Recall why the properties of polymers vary
- HT Describe how monomers join together in condensation polymerisation reactions.

Addition Polymerisation

- Because **alkenes** are **unsaturated**, they are useful for making other molecules, especially **polymers** (long-chain molecules).
- Many **monomers** (small molecules with double bonds) can join together to form polymers.
- This is called **addition polymerisation**.
- The materials commonly called plastics are all synthetic (man-made) polymers made in this way.

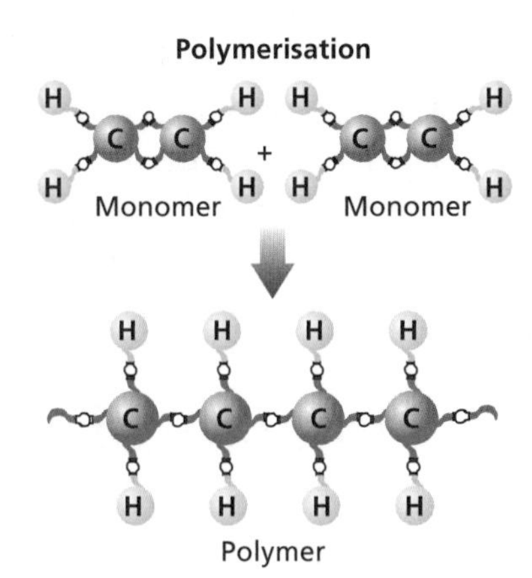

Representing Polymerisation

- Polymerisation can be represented like this:

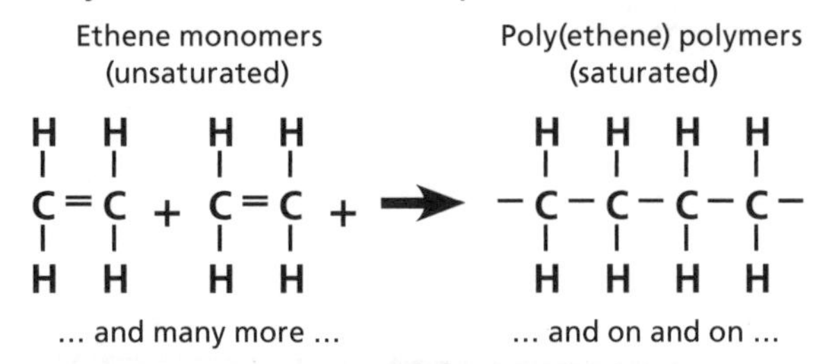

- The general formula for polymerisation can be used to represent the formation of any simple polymer:

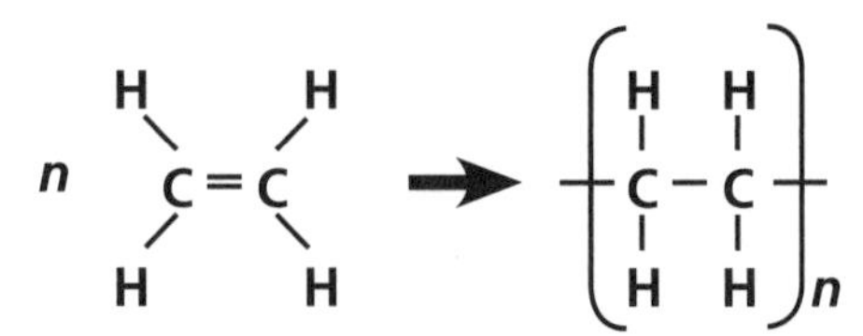

- In addition polymerisation reactions:
  - the repeating unit and the monomer units contain the same atoms
  - the percentage atom economy is 100%.

Properties of Polymers

- The properties of a polymer depend on:
  - what it is made from, i.e. what monomer was used
  - the conditions (e.g. temperature and catalyst) under which it was made.
- For example, low density poly(ethene) (LDPE) and high density poly(ethene) (HDPE) are both made from the monomer ethene.
- However, the polymers have different properties because different catalysts and reaction conditions are used to make them.
- LDPE is used to make carrier bags and HDPE is used to make plastic bottles.

Key Point

Polymers such as poly(ethene) and poly(propene) are made in addition polymerisation reactions.

Thermosoftening and Thermosetting Polymers

- Thermosoftening polymers such as poly(ethene):
  - consist of individual polymer chains that are tangled together (like spaghetti)
  - have weak intermolecular forces between all of the polymer chains and soften on heating.
- Thermosetting polymers such as melamine:
  - consist of polymer chains that are joined together by cross-links
  - do not melt when they are heated.

Thermosoftening Polymer

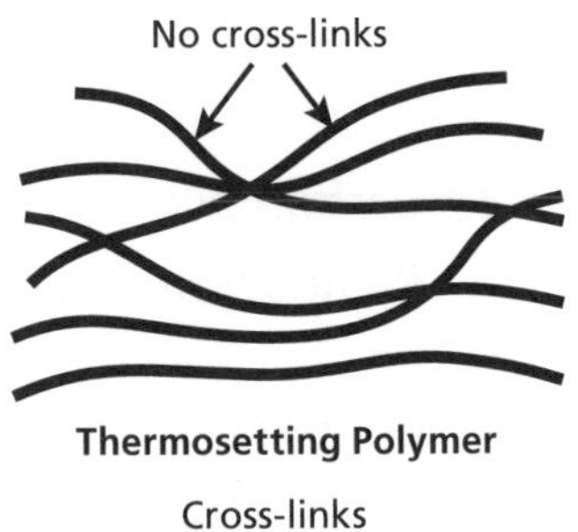

Thermosetting Polymer

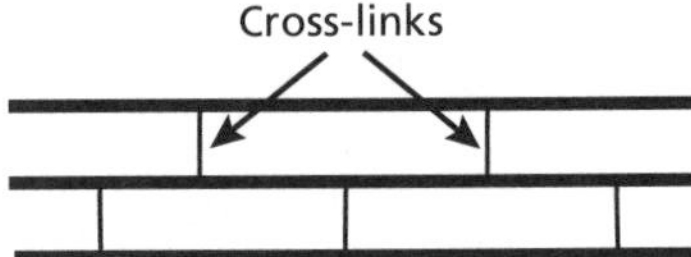

HT Condensation Polymerisation

- In **condensation polymerisation** reactions, monomer molecules join together to form large polymer molecules and lose small molecules such as water, H_2O, as by-products
- The simplest polymers are formed when **diols** (molecules with two hydroxyl, OH, groups) join together with **dicarboxylic acids** (molecules with two carboxyl, COOH, groups).
- For example, lots of ethanediol molecules react with lots of hexandoic acid molecules to form Terylene, a type of **polyester**:

$$n\ H{-}O{-}CH_2{-}CH_2{-}O{-}H + n\ HO{-}CO{-}CH_2{-}CH_2{-}CH_2{-}CH_2{-}CO{-}OH$$

$$\downarrow$$

$$\left[{-}CH_2{-}CH_2{-}O{-}CO{-}CH_2{-}CH_2{-}CH_2{-}CH_2{-}CO{-}O{-}\right]_n$$

$$+\ 2H_2O$$

- **Amino acids** join together by condensation polymerisation to form polypeptides and water.
- Polypeptides contain lots of peptide links.
- A peptide link is the bond formed between the carboxyl groups and the amino groups when amino acids join together.
- Glycine is an amino acid with the formula H_2NCH_2COOH.
- It polymerises to form a polypeptide with the formula $(\text{-}HNCH_2CO\text{-})_n$ and nH_2O.

Key Point

Polyesters are made by condensation polymerisation reactions.

Quick Test

1. Name the addition polymer formed when lots of styrene molecules join together.
2. What is the atom economy in addition polymerisation reactions?
3. HT What is a diol?
4. HT Name the type of reaction used to produce polyesters.
5. HT Name the bond between the carboxyl groups and the amino groups formed when amino acids join together.

Key Words

alkenes
unsaturated
polymer
monomer
addition polymerisation
HT **condensation polymerisation**
HT **diol**
HT **dicarboxylic acid**
HT **polyester**
HT **amino acid**

Review Questions

Conservation of Mass

1. HT Carbon dioxide is produced when carbon is burned in a good supply of oxygen.
Relative atomic masses (A_r): C = 12, O = 16

   a) Calculate the relative molecular mass of carbon dioxide, CO_2. [2]

   b) Calculate the mass of 1.00 mole of carbon dioxide, CO_2. [1]

   c) Calculate the mass of 2.00 moles of carbon dioxide, CO_2. [1]

2. HT Zinc is more reactive than iron.
Zinc displaces iron from a solution of iron sulfate solution:

$$Zn + FeSO_4 \rightarrow ZnSO_4 + Fe$$

   Complete the two half equations for this reaction:

   a) .... $\rightarrow Zn^{2+} + 2e^-$ [1]

   b) Fe^{2+} + ... $\rightarrow$ .... [2]

3. When calcium carbonate is heated fiercely, it decomposes to form calcium oxide and carbon dioxide.

$$CaCO_3(s) \rightarrow CaO(s) + CO_2(g)$$

   a) Why is the total mass of the reactants before the reaction equal to the total mass of reactants after the reaction? [1]

   b) What does the state symbol (s) mean? [1]

   c) 10.0g of calcium carbonate was heated until it had all reacted.
   5.6g of calcium oxide was produced.

   Why has the mass of the solid gone down? [2]

   d) Predict the mass of carbon dioxide produced in this reaction. [1]

4. Water vapour is produced when hydrogen is burned in a good supply of oxygen.
Relative atomic masses (A_r): H = 1, O = 16

   a) Calculate the relative formula mass of water, H_2O. [2]

   b) HT Calculate the mass of 0.50 moles of water, H_2O. [2]

5. Magnesium nitrate is an ionic compound.
Relative atomic masses (A_r): Mg = 24, N = 14, O = 16

a) Calculate the relative formula mass of magnesium nitrate, $Mg(NO_3)_2$. [2]

b) HT Calculate the mass of 1.00 mole of magnesium nitrate, $Mg(NO_3)_2$. [1]

6. A student places a piece of magnesium ribbon in a crucible.
They carefully heat the crucible and the magnesium.

$$2Mg(s) + O_2(g) \rightarrow 2MgO(s)$$

Why does the mass of the crucible go up? [2]

Total Marks .... / 21

HT Amount of Substance

1. What is the Avogadro constant?
Tick **one** box.

The number of particles in 1g of a substance. ☐

The number of particles in 1 mole of a substance. ☐

The number of protons in 1 mole of carbon. ☐

The number of subatomic particles in 1 mole of carbon. ☐ [1]

2. The Avogadro constant has a value of 6.02×10^{23}.

a) How many atoms are present in 23g of sodium? [1]

b) How many atoms are present in 6g of carbon? [1]

3. Calculate the number of moles in each of these substances:

a) 39g of potassium. [1]

b) 32g of sulfur dioxide, SO_2. [2]

c) 18g of ammonium ions, NH_4^+. [2]

4. Many fuels contain small amounts of sulfur.
When sulfur is burned, sulfur dioxide, SO_2, is produced

$$S + O_2 \rightarrow SO_2$$

1.6g of sulfur was completely burned in oxygen to produce sulfur dioxide.
Relative atomic masses (A_r): S = 32, O = 16

a) How many moles of sulfur were burned? [2]

b) Calculate the mass of sulfur dioxide, SO_2, produced in this reaction. [2]

5. Hydrogen was reacted with an excess of chlorine to produce hydrogen chloride.

$$H_2 + Cl_2 \rightarrow 2HCl$$

0.73g of hydrogen chloride was produced in this reaction.

a) Calculate the amount, in moles, of hydrogen chloride produced in this reaction. [2]

b) Calculate the amount, in moles, of hydrogen that reacted in this reaction. [2]

c) Calculate the mass of hydrogen that would produce 0.73g of hydrogen chloride. [2]

6. At room temperature and pressure, 1 mole of any gas takes up a volume of $24dm^3$.

a) Calculate the volume of 0.2 moles of nitrogen, N_2. [2]

b) Calculate the amount, in moles, of oxygen, O_2, gas present in $18dm^3$ at room temperature and pressure. [2]

Total Marks .... / 22

HT Titration

1. A student makes a solution of sodium hydroxide.
They place 1.00 mole of sodium hydroxide pellets in a volumetric flask.
They then add distilled water until the solution has a volume of $500cm^3$.
What is the concentration of this solution?
You must include the units in your answer. [2]

2. Titration can be used to measure how much alkali is needed to neutralise an acid.

25.0cm^3 of sodium hydroxide was placed in a flask.
The sodium hydroxide has a concentration of 0.1mol/dm^3.
This required 22.5cm^3 of hydrochloric acid solution for complete neutralisation.
The reaction can be summed up by the equation:

$$HCl + NaOH \rightarrow NaCl + H_2O$$

a) How many moles of sodium hydroxide were used in this reaction? [2]

b) How many moles of hydrochloric acid were used in this reaction? [1]

c) What is the concentration of the hydrochloric acid? [2]

Total Marks .... / 7

Percentage Yield and Atom Economy

1. The theoretical yield of a reaction is 15.0g of product.
A chemist carries out the reaction and produces only 9.0g of product.

Calculate the percentage yield of this reaction. [2]

2. Copper is a very useful metal.
It can be extracted from a solution of copper sulfate using scrap iron.
Relative formula mass (A_r): Fe = 56, Cu = 63.5, S = 32, O = 16

$$Fe + CuSO_4 \rightarrow Cu + FeSO_4$$

Reactions with a high atom economy contribute towards sustainable development.

a) Name the products of the reaction. [2]

b) Calculate the relative formula mass of iron sulfate, $FeSO_4$. [2]

c) Calculate the percentage atom economy of this reaction.
Give your answer to the nearest whole number. [2]

Total Marks .... / 8

Review Questions

Reactivity of Metals

1. Metals are usually extracted from their ores.

   a) What is an ore? [1]

   b) Explain how lead is extracted from lead oxide. [2]

Total Marks .... / 3

The pH Scale and Salts

1. Which of these ions is found in excess in alkaline solutions?
   Tick **one** box.

   H^+ ☐ H^- ☐ OH^+ ☐ OH^- ☐ [1]

2. HT Strong acids are completely ionised in water.

   a) Name a strong acid. [1]

   b) What does 'ionised' mean? [1]

3. Which of these pH values shows the pH of a strong acid?
   Tick **one** box.

   7 ☐ 14 ☐ 8 ☐ 1 ☐ [1]

4. The pH scale is used to measure the acidity or alkalinity of aqueous solutions.
   The pH of a solution can be found by using an indicator.

   a) How does an indicator work? [1]

   b) Suggest **one** other way that the pH of a solution could be found. [1]

   c) Hydrochloric acid is a strong acid.

   Complete the word equation to show the neutralisation reaction between hydrochloric acid and potassium hydroxide.

   hydrochloric acid + potassium hydroxide → .... + ... [2]

   d) What is the pH of a neutral solution? [1]

Total Marks .... / 9

Electrolysis

1 Lead bromide is an ionic compound that can be separated into lead and bromine.

a) Name the process used to separate the lead bromide. [1]

b) Lead bromide contains Pb^{2+} and Br^{-} ions.

Name the elements formed:

i) At the positive electrode. [1]

ii) At the negative electrode. [1]

c) Explain why the lead bromide must be molten for this process. [1]

d) Explain why the electrodes used in the experiment must be inert. [1]

e) HT Complete the half equations to show the reactions that take place at each electrode:

i) At the anode: $2Br^{-} \rightarrow Br_2 +$ .... [1]

ii) At the cathode: $Pb^{2+} + 2e^{-} \rightarrow$ .... [1]

2 Aluminium is extracted from its ore, aluminium oxide, by electrolysis.
Electrolysis is a very expensive process.

a) Why is electrolysis expensive? [1]

b) During the extraction of aluminium, the main ore, bauxite (which contains aluminium oxide), is mixed with cryolite (another ore of aluminium).

Why is the bauxite mixed with cryolite? [1]

c) Name the substance formed at the negative electrode during the extraction of aluminium. [1]

d) HT Complete the half equation for the reaction that takes place at the anode during the electrolysis of aluminium oxide.

$$2O^{2-} \rightarrow O_2 + \text{..........}$$ [1]

Total Marks .... / 11

Practice Questions

Exothermic and Endothermic Reactions

1 Which of the following types of reaction is an endothermic reaction?
Tick **one** box.

Combustion ☐ Oxidation ☐

Neutralisation ☐ Thermal decomposition ☐ [1]

2 Complete the energy profile diagram in **Figure 1** to show an exothermic reaction.

Include the:
- Reactants
- Products
- Activation energy
- Energy change of the reaction [4]

Figure 1

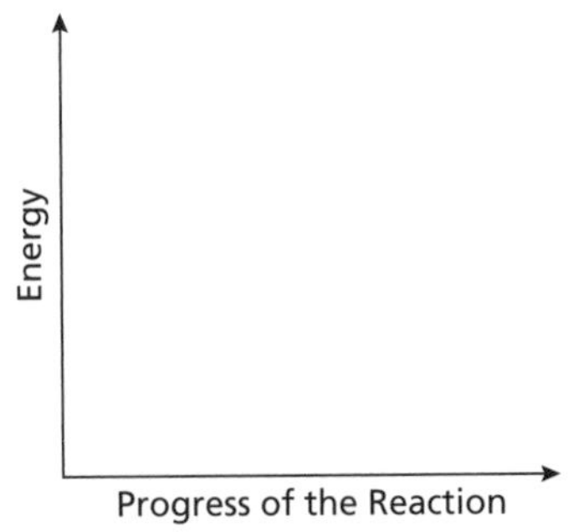

Energy

Progress of the Reaction

3 A student carries out an experiment to find out whether changing the metal powder added to dilute hydrochloric acid affects the temperature change for the reaction.

a) Why must the student take the temperature at the start and the end of each reaction? [1]

b) Why must the student stir the mixture of metal powder and acid? [1]

c) Name a control variable in this experiment. [1]

d) Name a hazard in this experiment and describe a control measure the student must take to reduce the risk of an accident happening. [2]

Total Marks .... / 10

Fuel Cells

1 Traditionally electricity has been produced by burning fossil fuels in power stations. Hydrogen fuel cells are a more efficient way of producing electricity than traditional methods.

Give **three** other advantages of using fuel cells over traditional methods of producing electricity. [3]

Total Marks .... / 3

Rate of Reaction

1 Which of the following changes would increase the rate of a chemical change?
Tick **one** box.

Adding a catalyst	☐	Increasing the size of particles	☐
Reducing the temperature by 10°C	☐	Decreasing the concentration	☐

[1]

2 Collision theory can be used to explain the rate of a chemical reaction.

a) What **two** things must happen for a chemical reaction to take place? [2]

b) Explain, in terms of collision theory, why increasing the temperature increases the rate of a chemical change. [2]

3 In the 'disappearing cross' experiment, hydrochloric acid reacts with sodium thiosulfate. One of the products of the reaction is sulfur, which is insoluble.
A student carries out the experiment using 1.00mol/dm^3 acid.
She adds the acid to the sodium thiosulfate and times how long it takes for the cross to 'disappear'.
She then repeats the experiment using 0.500mol/dm^3 and then 0.250mol/dm^3 hydrochloric acid.

Timer
Add dilute hydrochloric acid
Flask
Sodium thiosulfate
Paper with cross drawn on it

a) Name the independent variable in this investigation. [1]

b) Temperature could affect the rate of reaction, so the same temperature must be used for each part of the investigation.

Predict and explain how increasing the temperature would affect the time taken for the cross to disappear. [2]

Total Marks .... / 8

Reversible Reactions

1 If a reversible reaction is carried out in a closed system, equilibrium can be reached.

a) What is a closed system? [1]

b) Explain how you can tell that a system is in equilibrium. [1]

Total Marks .... / 2

Practice Questions

Alkanes

1. Ethane is used as a fuel.

   a) What is a fuel? [1]

   b) Complete the equation for the complete combustion of ethane.

   $$2C_2H_6 + \dots O_2 \rightarrow \dots CO_2 + \dots H_2O$$ [3]

   c) Carbon monoxide and water vapour are produced in the **incomplete** combustion of ethane.

   Name **one** other product of the incomplete combustion of ethane. [1]

2. Crude oil is a mixture of different substances.

   Describe how crude oil is formed. [3]

3. Fractional distillation of crude oil produces a large amount of long-chain hydrocarbons.

   a) Which of these is a property of long-chain hydrocarbons?
   Tick **one** box.

Viscous	☐	Very flammable	☐
Low melting points	☐	Runny	☐

   [1]

   b) There is a large supply but low demand for long-chain hydrocarbons.
   In industry, long-chain hydrocarbons are cracked to produce more useful products.

   Complete the equation for the cracking of this long carbon chain hydrocarbon.

   $$C_{20}H_{42} \rightarrow C_{18}H_{38} + \dots$$ [1]

4. Methane, CH_4, is an alkane.
   Alkanes are saturated hydrocarbons.

   a) What is a saturated hydrocarbon? [2]

   b) Give the general formula for alkanes. [1]

   c) Give the formula for the fourth member of the alkane homologous series. [1]

5 Name the type of bonds found in alkane molecules.
Tick **one** box.

Metallic ☐ Single covalent ☐

Ionic ☐ Double covalent ☐ [1]

6 Liquefied petroleum gas (LPG) contains propane and butane.

a) Give the formula of:

i) Propane. [1]

ii) Butane. [1]

b) Give **one** use of LPG. [1]

Total Marks .... / 18

Alkenes

1 Which of the substances below is an unsaturated hydrocarbon?
Tick **one** box.

C_2H_4 ☐ C_2H_6 ☐

CH_4 ☐ C_3H_8 ☐ [1]

2 Bromine water can be used to identify alkenes, such as propene.

a) What is the formula of propene? [1]

b) The bromine water is added to a sample and the mixture is shaken.

Which of these observations would show that propene was present?
Tick **one** box.

White precipitate ☐ No colour change ☐

Cream precipitate ☐ Orange / brown to colourless ☐ [1]

Total Marks .... / 3

Practice Questions

Organic Compounds

1. Esters are produced when carboxylic acids are warmed with alcohols.

   Name the ester produced when ethanol reacts with ethanoic acid. [1]

2. An organic compound has the formula CH_3CH_2COOH.

   What type of compound is it?
   Tick **one** box.

Alkane	☐	Ester	☐
Alcohol	☐	Carboxylic acid	☐

   [1]

3. Ethanol can be produced by reacting ethene with steam in a hydration reaction.

   a) Complete the equation to show the hydration of ethene.

   $$C_2H_4 + \ldots\ldots \rightarrow \ldots\ldots$$ [2]

   b) Why is this method of producing ethanol non-renewable? [1]

   c) What is the atom economy for the hydration of ethene?
   You must give a reason for your answer. [2]

4. HT Glycine is an amino acid.

   a) Name the **two** functional groups found in amino acids. [2]

   b) Different amino acids can join together to form important new molecules.
   What sorts of molecules are made when different amino acids join together? [1]

5. An organic compound has the formula $CH_3CH_2CH_2OH$.

   a) What is the functional group of this organic compound? [1]

   b) Name this organic compound. [1]

   c) Suggest the pH of an aqueous solution of this organic compound. [1]

6. Ethanol can be produced by the fermentation of glucose.

   a) What is the catalyst used in this reaction? [1]

   b) A temperature of 25°C to 50°C is usually used for fermentation.
   If too high or too low a temperature is used, the rate of reaction can change.

What would happen to the catalyst if the temperature was:

i) Too high? [1]

ii) Too low? [1]

7 HT DNA stores and transmits the instructions for the development of living organisms. DNA is made of two polymer chains.

a) How many different nucleotides is DNA made from?
Tick **one** box.

2 ☐ 4 ☐ 23 ☐ 46 ☐ [1]

b) DNA is made up of two strands.

What two words describe the structure of DNA? [1]

Total Marks .... / 18

Polymerisation

1 Ethene can be used to make polymers.

a) Name the polymer produced when lots of ethene molecules are joined together. [1]

b) What name is given to a reaction in which two or more molecules join together to give a single product? [1]

c) What is the percentage atom economy of this reaction? [1]

2 Diol and dicarboxylic acid molecules can be reacted together to form a polymer.

a) What is a 'diol'? [1]

b) What type of polymer is produced when diol molecules react with dicarboxylic acid molecules? [1]

c) Draw a diagram to show the functional group in the polymer produced in part **b)**. [1]

3 Name the type of reaction that produces polyester from diols and dicarboxylic acids.
Tick **one** box.

Fractional distillation ☐

Cracking ☐

Addition polymerisation ☐

Condensation polymerisation ☐ [1]

Total Marks .... / 7

Chemical Analysis

You must be able to:

- Explain what the term 'pure' means in chemistry
- Explain what a formulation is
- Explain how chromatography can be used to separate mixtures
- Identify a range of gases
- Understand why modern instrumental methods of analysis are useful.

Pure and Impure Substances

- In chemistry, the word **pure** has a special meaning – a pure substance contains only one type of element or one type of compound.
- This means that pure substances:
  - melt and solidify at one temperature called the melting point
  - boil and condense at one temperature called the boiling point.
- Impure substances are mixtures. They do not melt and boil at one temperature – they change state over a range of temperatures.

Key Point

A pure substance contains only one type of element or one type of compound.

Formulations

- **Formulations** are mixtures that have been carefully designed to have specific properties.
- The components in a formulation are carefully controlled.
- Examples of formulations include fuels, cleaning agents, paints, medicines, alloys, fertilisers and foods.

Chromatography

- **Chromatography** involves:
  - a **stationary phase**, which does not move
  - a **mobile phase**, which does move.
- In paper chromatography:
  - the stationary phase is the absorbent paper
  - the mobile phase is the solvent, which is often water.
- During chromatography, mixtures are separated into their constituent components.
- The solvent dissolves the samples and carries them up the paper.
- Each component moves a different distance up the paper depending on its attraction for the paper and for the solvent.
- Chromatography can be used to identify artificial colours (e.g. in food) by comparing them to the results obtained from known substances.

REQUIRED PRACTICAL

Investigate how paper chromatography can be used to separate and tell the difference between coloured substances.

Sample Method	Considerations, Mistakes and Errors
1. Draw a 'start line', in pencil, on a piece of absorbent paper. 2. Put samples of five known food colourings (A, B, C, D and E), and the unknown substance (X), on the 'start line'. 3. Dip the paper into a solvent. 4. Wait for the solvent to travel to the top of the paper. 5. Identify substance X by comparing the horizontal spots with the results of A, B, C, D and E.	• Pure substances produce a single spot in all solvents. • Only ever use pencil to draw the start line, as ink will dissolve and affect your results.

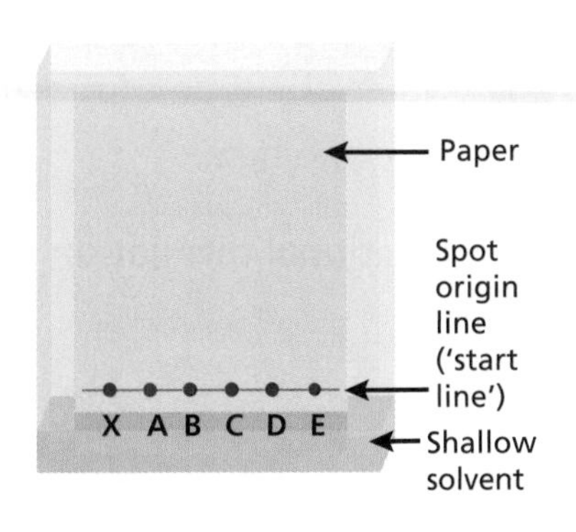

- R_f values can be used to identify the components in a mixture.

$$R_f = \frac{\text{distance moved by substance}}{\text{distance moved by solvent}}$$

- Different components have different R_f values.
- Providing the same temperature and solvent are used, the R_f value for a particular component is constant and can be used to identify the component.

Gas Tests

Gas	Properties	Test for Gas
Hydrogen, H_2	A colourless gas. It combines violently with oxygen when ignited.	When mixed with air, burns with a squeaky pop.
Chlorine, Cl_2	A green poisonous gas that bleaches dyes.	Turns damp indicator paper white.
Oxygen, O_2	A colourless gas that helps fuels burn more readily than in air.	Relights a glowing splint.
Carbon dioxide, CO_2	A colourless gas.	When bubbled through limewater (a solution of calcium hydroxide), turns the limewater cloudy.

Instrumental Methods

- Standard laboratory equipment can be used to detect and identify elements and compounds.
- However, methods that involve using highly accurate instruments to analyse and identify substances have been developed to perform this function more effectively.
- **Flame emission spectroscopy** is a very useful instrumental method.
- It is used to analyse solutions that contain metal ions.
- A sample of the metal solution is placed in a flame and the light emitted is passed through a spectroscope.
- This produces a line spectrum which can be used to:
  - identify the metal ions in the solution
  - measure the concentration of the metal ions.

Key Point

Modern instrumental methods of detection and analysis give rapid results, are very sensitive and accurate, and can be used on small samples.

Quick Test

1. In chemistry what does the term 'pure' mean?
2. In paper chromatography what is the:
   a) Stationary phase?
   b) Mobile phase?
3. Give **four** advantages of instrumental methods of analysis.
4. What type of substances cam be analysed using flame emission spectroscopy?

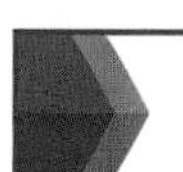

Key Words

pure
formulation
chromatography
stationary phase
mobile phase
flame emission spectroscopy

Identifying Substances

You must be able to:

- Explain how to identify metal ions from flame tests
- Describe how to identify metal carbonates
- Describe how to identify metal ions in solutions using sodium hydroxide solution
- Describe how sulfates and halide ions can be identified.

Flame Tests

- **Flame tests** can be used to identify metal ions (cations).
- Lithium, sodium, potassium, calcium and copper compounds can be recognised by the distinctive colours they produce in a flame test.

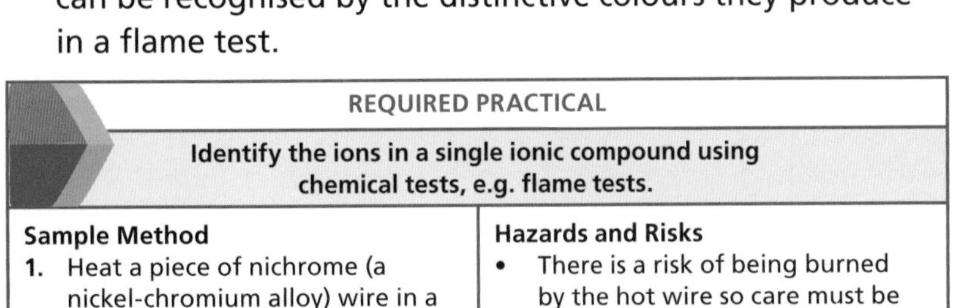

REQUIRED PRACTICAL

Identify the ions in a single ionic compound using chemical tests, e.g. flame tests.

Sample Method	Hazards and Risks
1. Heat a piece of nichrome (a nickel-chromium alloy) wire in a Bunsen flame and then dip it in concentrated hydrochloric acid to clean it. 2. Dip the wire in the compound. 3. Put it into a Bunsen flame and observe what colour flame is produced.	• There is a risk of being burned by the hot wire so care must be taken not to touch it. • The concentrated acid is corrosive so avoid skin contact and use eye protection. • The compounds used may be harmful so avoid skin contact.

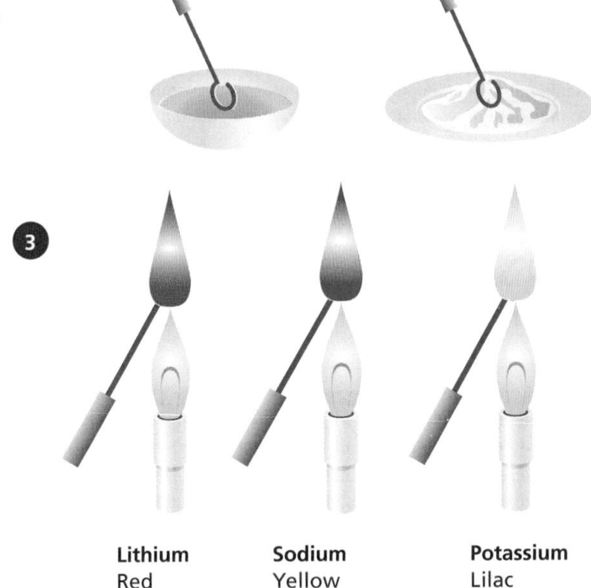

- The following distinctive colours indicate the presence of certain ions:
  - green for copper
  - brick red for calcium
  - crimson red for lithium
  - lilac for potassium
  - yellow for sodium.
- If a sample contains a mixture of ions, the colours of some ions can be masked.
- For example if a solution contains sodium and potassium ions, the pale lilac colour from the potassium ions can be hard to detect alongside the intense yellow from the sodium ions.

Key Point

You need to be able to carry out all the chemical tests in this topic.

Reacting Carbonates with Dilute Acids

- Carbonates react with dilute acids to form carbon dioxide gas (plus a salt and water), e.g.

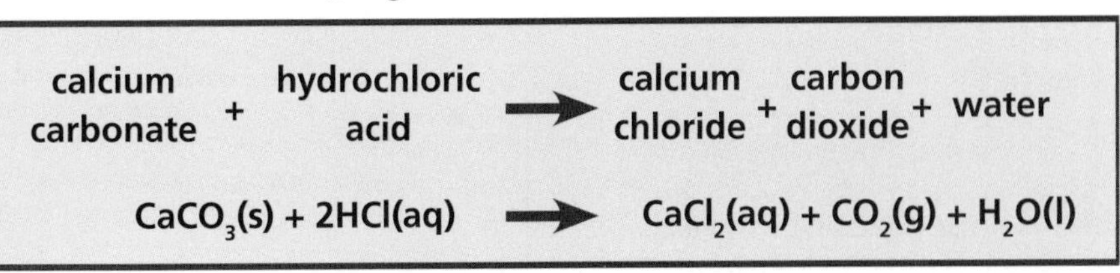

calcium carbonate + hydrochloric acid ⟶ calcium chloride + carbon dioxide + water

$$CaCO_3(s) + 2HCl(aq) \longrightarrow CaCl_2(aq) + CO_2(g) + H_2O(l)$$

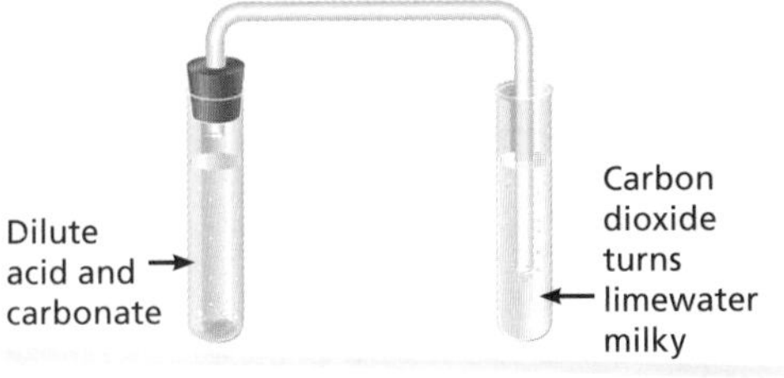

- Carbon dioxide turns limewater cloudy (milky).
- Most metal carbonates are insoluble.

- However, sodium carbonate and potassium carbonate are soluble and dissolve in water to form solutions that contain carbonate ions.

Precipitation of Metal Ions

- Solutions of metal compounds contain metal ions.
- Some of these form **precipitates**, i.e. **insoluble** solids that come out of solution, when sodium hydroxide solution is added to them.
- For example, when sodium hydroxide solution is added to calcium chloride solution, a white precipitate of calcium hydroxide is formed (as well as sodium chloride solution).

HT You can see how this precipitate is formed by looking at the ions involved:

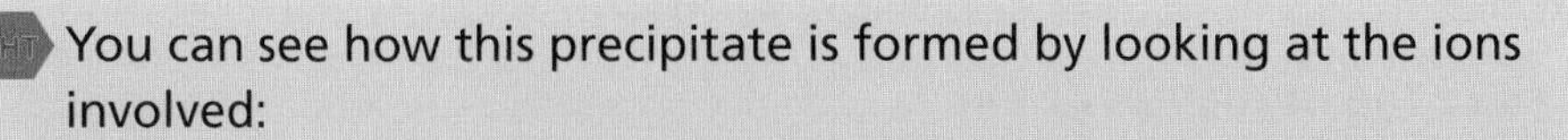

$$Ca^{2+}(aq) + 2OH^{-}(aq) \longrightarrow Ca(OH)_2(s)$$

- The table below shows the precipitates formed when metal ions are mixed with sodium hydroxide solution:

Key Point

Some metal ions can be identified by adding sodium hydroxide solution to solutions of metal ions to form precipitates.

Metal Ion	Precipitate Formed	Precipitate Colour
aluminium, $Al^{3+}(aq)$	aluminium hydroxide	white – dissolves if more sodium hydroxide solution is added
calcium, $Ca^{2+}(aq)$	calcium hydroxide	white
magnesium, $Mg^{2+}(aq)$	magnesium hydroxide	white
copper(II), $Cu^{2+}(aq)$	copper(II) hydroxide	blue
iron(II), $Fe^{2+}(aq)$	iron(II) hydroxide	green
iron(III), $Fe^{3+}(aq)$	iron(III) hydroxide	brown

Sulfates and Halides

- If dilute hydrochloric acid and barium chloride solution are added to a solution containing sulfate ions, a white precipitate of barium sulfate is produced.
- Solutions of **halide** ions (chloride, bromide and iodide ions) react with silver nitrate solution in the presence of dilute nitric acid to produce silver halide precipitates:
  - silver chloride is white
  - silver bromide is cream
  - silver iodide is yellow.

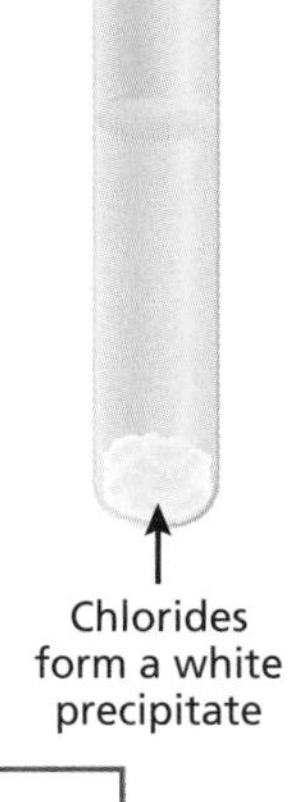

Chlorides form a white precipitate

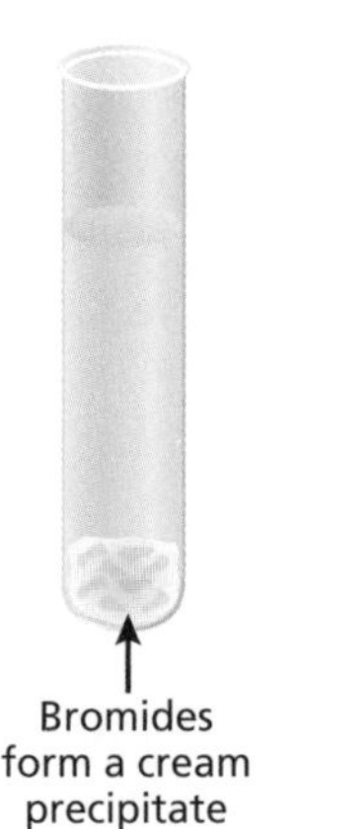

Bromides form a cream precipitate

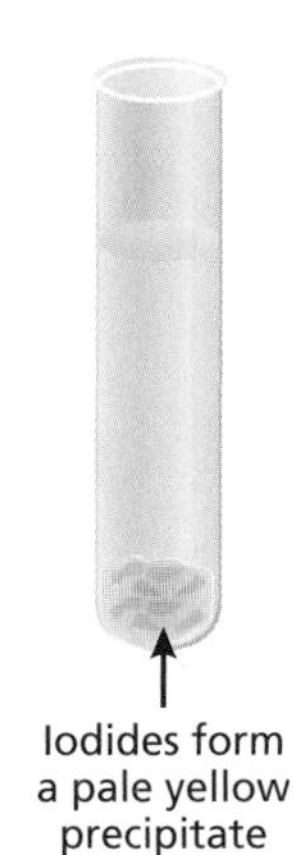

Iodides form a pale yellow precipitate

Quick Test

1. What colour flame is produced when a sample of copper chloride is used in a flame test?
2. What is a precipitate?
3. What colour is a copper(II) hydroxide precipitate?
4. How can a sulfate be identified?
5. What colour is a precipitate of silver bromide?

Key Words

flame tests
precipitate
insoluble
halide

The Earth's Atmosphere

You must be able to:

- Recall the present day composition of the Earth's atmosphere
- Describe how and why the Earth's atmosphere has changed over time
- Explain why the levels of oxygen have increased over time
- Explain why the levels of carbon dioxide have decreased over time.

The Earth's Atmosphere

- The atmosphere has changed a lot since the formation of the Earth 4.6 **billion** years ago.

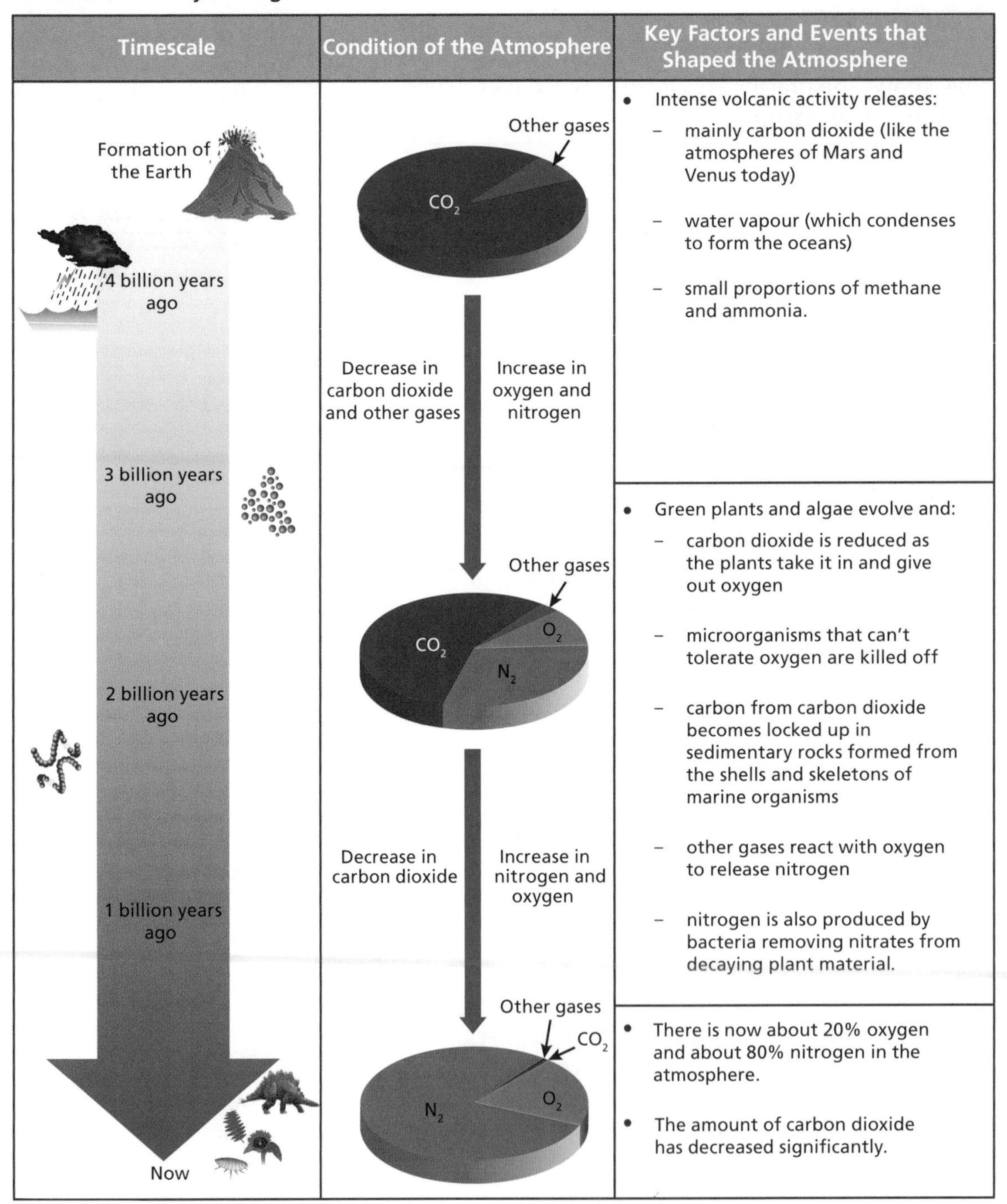

Timescale	Condition of the Atmosphere	Key Factors and Events that Shaped the Atmosphere
Formation of the Earth 4 billion years ago	Pie chart: CO_2, Other gases Decrease in carbon dioxide and other gases / Increase in oxygen and nitrogen	• Intense volcanic activity releases: – mainly carbon dioxide (like the atmospheres of Mars and Venus today) – water vapour (which condenses to form the oceans) – small proportions of methane and ammonia.
3 billion years ago 2 billion years ago 1 billion years ago	Pie chart: CO_2, N_2, O_2, Other gases Decrease in carbon dioxide / Increase in nitrogen and oxygen	• Green plants and algae evolve and: – carbon dioxide is reduced as the plants take it in and give out oxygen – microorganisms that can't tolerate oxygen are killed off – carbon from carbon dioxide becomes locked up in sedimentary rocks formed from the shells and skeletons of marine organisms – other gases react with oxygen to release nitrogen – nitrogen is also produced by bacteria removing nitrates from decaying plant material.
Now	Pie chart: N_2, O_2, CO_2, Other gases	• There is now about 20% oxygen and about 80% nitrogen in the atmosphere. • The amount of carbon dioxide has decreased significantly.

The Atmosphere Today

- The proportions of gases in the atmosphere have been more or less the same for about 200 million years.
- Water vapour may also be present in varying quantities (0–3%).

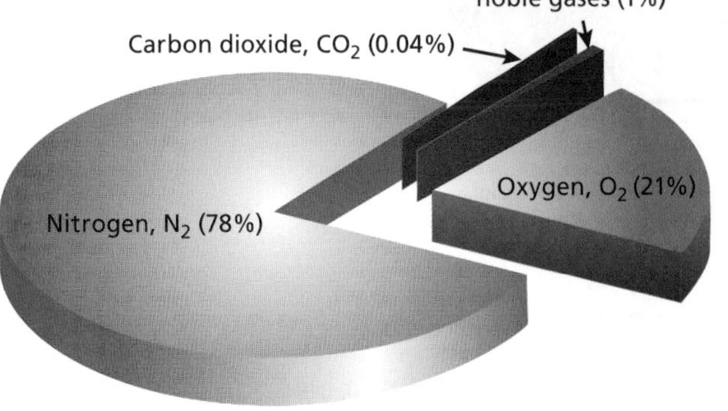

Increase of Oxygen Levels

- **Algae** and plants photosynthesise.
- During **photosynthesis**, carbon dioxide and water react to produce glucose and oxygen:

> **carbon dioxide + water → glucose + oxygen**
> $$6CO_2 + 6H_2O \rightarrow C_6H_{12}O_6 + 6O_2$$

- Algae first started producing oxygen about 2.7 billion years ago.
- Over the next billion years, plants evolved and the amount of oxygen in the atmosphere increased.
- Eventually the level of oxygen in the atmosphere increased enough to allow animals to evolve.

Key Point

The Earth's early atmosphere was mainly carbon dioxide. Over time the levels of oxygen have increased and the levels of carbon dioxide have decreased.

Decrease of Carbon Dioxide Levels

- As plants and algae have evolved, the level of carbon dioxide in the atmosphere has decreased. This is because plants use carbon dioxide during photosynthesis.
- Carbon dioxide has also decreased as carbon becomes locked up in sedimentary rocks (e.g. limestone) and fossil fuels (e.g. coal, crude oil and natural gas).
- Limestone contains calcium carbonate and can be formed from the shells and skeletons of sea creatures.
- Coal is a sedimentary rock formed from plant deposits that were buried and compressed over millions of years.
- The level of carbon dioxide in the atmosphere has also been reduced by the reaction between carbon dioxide and sea water. This reaction produces:
  - insoluble carbonates that are deposited as sediment
  - soluble hydrogen carbonates.
- However, too much carbon dioxide dissolving in the oceans can harm marine life, such as coral reefs.

Quick Test

1. What is the main gas in the Earth's atmosphere today?
2. Name the main gas in the Earth's early atmosphere.
3. The atmospheres of which planets are thought to be like the Earth's early atmosphere?
4. What is the main compound in limestone rocks?
5. Write the balanced symbol equation for photosynthesis.

Key Words

billion
algae
photosynthesis

Greenhouse Gases

You must be able to:

- Recall the names of some greenhouse gases
- Explain how greenhouse gases increase the Earth's temperature
- Describe how human activities can increase the levels of greenhouse gases in the atmosphere
- Describe the effects of global climate change
- Understand the factors that contribute to and affect a carbon footprint.

Greenhouse Gases

- High energy, short wavelength **infrared** radiation from the Sun passes through the atmosphere and reaches the Earth's surface.
- Some of this radiation is absorbed by the Earth.
- However, lower energy, longer wavelength infrared radiation is reflected by the Earth's surface.
- **Greenhouse gases** in the Earth's atmosphere absorb this outgoing infrared radiation, which increases the Earth's temperature.
- Without some greenhouse gases, the Earth would be too cold for water to be a liquid and would not be able to support life.
- Greenhouse gases include carbon dioxide, water vapour and methane.

The Impact of Human Activities

- Human activities can increase the levels of greenhouse gases in the atmosphere.
- The amount of carbon dioxide in the atmosphere has increased over the last 100 years. This increase correlates with an increase in the amount of fossil fuels being burned.
- Fossil fuels contain carbon that has been locked up for millions of years.
- Burning them releases carbon dioxide into the atmosphere.
- Deforestation also leads to an increase in carbon dioxide in the atmosphere as there are fewer trees taking up the gas for photosynthesis.
- Activities that increase the levels of methane in the atmosphere include:
  - the decomposition of rubbish in landfill sites
  - the increase in animal farming – it is produced by animals during digestion and by the decomposition of their waste materials.
- Many scientists believe that the increase in the levels of greenhouse gases in the atmosphere will increase the temperature of the Earth's surface and could result in global climate change.
- However, with so many different factors involved, it is difficult to produce an accurate model for such a complicated system.
- As a result, people may use simplified models.
- This can lead to speculation and opinions being expressed to the media that may be based on only part of the evidence.
- In addition, some people may have views that are **biased**, e.g. people being paid by companies that produce greenhouse gases and who have a vested interest in these issues.

Key Point

Climate change is a good example of the popular media reporting on scientific ideas in a way that may be oversimplified, inaccurate or biased.

It is also a good example of an area where scientists can work to tackle problems caused by human impacts on the environment.

Global Climate Change

- If the average global temperature increases this could cause global climate change. The impact of this could include:
  - a rise in sea level, which could cause devastating floods and more coastal erosion
  - more frequent and severe storm events
  - changes in the amount and timing of rainfall, with some areas receiving more rain and other areas receiving much less
  - an increased number of heatwave events, which can be harmful to people and wildlife
  - more droughts
  - changes to the distribution of plants and animals, as some areas become too hot for species to survive and other areas warm up enough to become habitable
  - food shortages in some areas, due to changes in the amount of food that countries can produce.

Key Point

Climate change is an area where governments and individuals need to make decisions about the best course of action by evaluating evidence and considering all arguments.

Carbon Footprints

- The **carbon footprint** of a product, service or event is the total amount of carbon dioxide and other greenhouse gases, such as methane, that are emitted over its full life cycle.
- For a product, this includes the production, use and disposal of the item.
- The carbon footprint can be reduced through:
  - using more alternative energy supplies, e.g. solar power
  - wasting less energy
  - carbon capture and storage (CCS), to prevent carbon dioxide being released into the atmosphere
  - carbon taxes and licences, to deter companies and individuals from choosing options that release lots of greenhouse gas
  - carbon off-setting, through activities such as tree planting
  - encouraging people to choose carbon-neutral products.
- However, reducing the carbon footprint is not straightforward.
- Problems include:
  - disagreement between scientists over the causes and consequences of global climate change
  - lack of information and knowledge in the general population
  - the reluctance of people to make lifestyle changes
  - economic considerations, such as the high cost of producing electricity from alternative energy resources rather than using cheaper fossil fuels
  - disagreement between countries as to what should be done.

Key Point

Trees use carbon dioxide for photosynthesis and, therefore, reduce the net amount of carbon dioxide reaching the atmosphere. The idea of carbon off-setting is to plant enough trees to balance out the carbon dioxide being produced by manufacturing processes / product use.

Key Point

Carbon-neutral products lead to no overall increase in the amount of carbon dioxide in the atmosphere.

Quick Test

1. Name **three** greenhouse gases.
2. Explain why, without greenhouse gases, the Earth could not support life.
3. What human activities are increasing the levels of carbon dioxide in the atmosphere?
4. What human activities are increasing the levels of methane in the atmosphere?
5. What does CCS stand for?

Key Words

infrared
greenhouse gases
biased
carbon footprint

Earth's Resources

You must be able to:

- Recall the resources that humans need to survive
- Understand how chemists can contribute towards sustainable development
- Describe how drinking water is produced
- Describe how waste water is treated
- HT Understand how and why copper is extracted from low-grade ores.

Sustainable Development

- Humans rely on the Earth's resources to provide them with warmth, shelter, food and transport.
- All our resources come from the Earth's crust, oceans or atmosphere.
- These resources can be renewable, such as timber, or finite (non-renewable), such as metal ores.
- Finite resources must be used with great care.
- Care must also be taken to ensure that the planet does not become too polluted.
- In the past, natural resources were sufficient to provide the human population with food, timber, clothing and fuels.
- However, as the population has increased, humans have come to rely on **agriculture** to supplement or even replace such resources.
- Chemistry plays an important role in improving agricultural and industrial processes – allowing new products to be developed and contributing towards **sustainable development**.

Key Point

Sustainable development meets the needs of the current generation without compromising the ability of future generations to meet their own needs.

Drinking Water

- Water of the correct quality is essential for life.
- Water naturally contains microorganisms and dissolved salts.
- These need to be at low levels for the water to be safe for humans to drink.
- **Fresh water** contains low levels of dissolved salts.
- Water that is good quality and safe to drink is called **potable**.
- In the UK potable water is produced in the following way:
  1. Fresh water from a suitable source, e.g. a lake or river away from polluted areas, is collected.
  2. It is passed through a filter bed to remove solid particles.
  3. Chlorine gas is added to kill any harmful microorganisms.
  4. Fluoride is added to drinking water to reduce tooth decay (although too much fluoride can cause discolouration of teeth).
- Ozone and ultraviolet can also be used sterilise water.
- To improve the taste and quality of tap water, more dissolved substances can be removed by passing the water through a filter containing carbon, silver and ion exchange resins.
- If fresh water supplies are limited, seawater can be **desalinated** to produce pure water. This can be done by distillation or reverse osmosis.
- Both of these processes use a lot of energy, making them very expensive.
- During distillation:
  - the water is boiled to produce steam
  - the steam is condensed to produce pure liquid water.

Waste Water Treatment

- Large amounts of waste water are produced by homes, agricultural processes and industrial processes.
- This waste water must be treated before it can be safely released back into the environment.
- Organic matter, harmful microorganisms and toxic chemicals have to be removed from sewage and agricultural and industrial waste water.
- Sewage treatment includes:
  - screening and grit removal
  - sedimentation to produce sewage sludge and effluent
  - anaerobic digestion of sewage sludge
  - aerobic biological treatment of effluent.

Key Point

Pure water contains no dissolved substances.

HT Alternative Methods of Extracting Metals

- Copper is a useful metal because:
  - it is a good conductor of electricity and heat
  - it is easily bent, yet hard enough to make water pipes and tanks
  - it does not react with water, so lasts for a long time.
- Copper can be extracted from copper-rich ores by heating the ores with carbon in a furnace. This process is known as smelting.
- The copper can then be purified by electrolysis.
- Copper can also be obtained:
  - from solutions of copper salts by electrolysis
  - by displacement using scrap iron.
- During electrolysis the positive copper ions move towards the negative electrode and form pure copper.
- The extensive mining of copper in the past means that we are running out of copper-rich ores.
- As a result, new methods have been developed to extract it from ores that contain less copper.
- Copper can be extracted from:
  - low-grade ores (ores that contain small amounts of copper)
  - contaminated land by biological methods.
- **Phytomining** is a method that uses plants to absorb copper:
  - As the plants grow they absorb (and store) copper.
  - The plants are then burned and the ash produced contains copper in relatively high quantities.
- **Bioleaching** uses bacteria to extract metals from low-grade ores:
  - A solution containing bacteria is mixed with a low-grade ore.
  - The bacteria convert the copper into a solution (known as a leachate solution), from which copper can be easily extracted.

Key Point

Phytomining and bioleaching are more environmentally friendly than traditional mining methods, which involve digging up and moving large quantities of rock and having to dispose of large amounts of waste materials.

Quick Test

1. Why is chlorine added to drinking water?
2. HT What is phytomining?
3. HT What is bioleaching?

Key Words

agriculture
sustainable development
fresh water
potable
desalinated
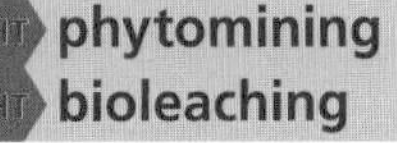
HT phytomining
HT bioleaching

Using Resources

You must be able to:

- Describe and explain different methods used to protect iron and steel objects from corrosion
- Explain what the life cycle assessment, LCA, of a product is
- Explain how an LCA can be used to help people make good decisions about which product to buy.

Preventing Corrosion

- Metals **corrode** when they react with oxygen and water in the environment.
- The term **rusting** is often used when iron objects, or objects containing iron (e.g. steel), corrode:

iron + water + oxygen ⟶ hydrated iron(III) oxide

- Corrosion can be prevented in several different ways.
- Painting, electroplating or greasing a metal object stop oxygen and water from reaching the surface and, therefore, stop corrosion from occurring.
- However, if the coating is damaged the metal will start to corrode.
- In **sacrificial protection**, a more reactive metal is placed in contact with the metal. For example:
  - Blocks of the more reactive magnesium metal are attached to the iron or steel object.
  - As magnesium is more reactive than iron, it reacts and loses electrons instead of the iron.
- **Galvanising** can also be used to protect iron or steel objects:
  - The object is coated in a layer of zinc.
  - The zinc layer stops oxygen and water from reaching the metal and stops corrosion.
  - If the surface of the zinc is scratched, it does not matter – the zinc provides sacrificial protection.
- Aluminium objects are protected from corrosion by a thin layer of aluminium oxide, which acts as a barrier and prevents the aluminium from reacting further.

Key Point

Sacrificial protection is so called because the more reactive metal reacts and loses its electrons instead of the iron or steel object that it is protecting.

Glass, Ceramics and Composites

- Glass is a non-crystalline solid and there are many different types.
- Soda-lime glass is made by heating a mixture of sand, sodium carbonate and limestone and is used as window glass.
- Borosilicate glass, or Pyrex, is made by heating sand and boron trioxide to a higher temperature than that used for producing soda-lime glass.
- Borosilicate glass is used to make chemical glassware, cooking equipment and car headlights.
- Pottery and bricks are clay ceramics. They are made by shaping wet clay and then heating them in a furnace.
- In the furnace, water is removed and chemical reactions take place that make the object retain its shape and become harder and stronger.

- **Composite materials** consist of two materials with different properties.
- The materials are combined together to produce a material that has its own improved properties.
- Examples of composites include concrete and fibreglass.

Life Cycle Assessment (LCA)

- A **life cycle assessment (LCA)** is used to assess the environmental impact a product has over its whole lifetime.
- They provide a way of comparing several alternative products to see which one causes the least damage to the environment.
- To carry out an LCA, scientists measure the impact of:
  - extracting the raw materials
  - processing the raw materials
  - manufacturing the product
  - how the product is used
  - how the product is transported
  - how the product is disposed of at the end of its life.
- Some aspects of the LCA are quite easy to quantify, e.g. the amounts of energy, water and raw materials used.
- However, some aspects of the LCA are difficult to quantify and involve value judgements, e.g. the impact of a pollutant on the environment, meaning a LCA is not completely objective.

Reducing the Use of Resources

- Materials such as glass, metals and plastics are important to our standard of living. However, they must be used wisely and reused and recycled wherever possible to:
  - save money and energy
  - make sure natural resources are not used up unnecessarily
  - reduce the amount of waste produced
  - reduce damage to the environment caused by extraction.
- Metal, glass, building materials and plastics made from crude oil are produced from limited resources.
- Our supplies of these raw materials, and the fossil fuels often used to obtain them, are finite.
- The mining and quarrying processes used to extract these raw materials can have devastating environmental impacts.
- Some objects such as plastic bags and glass bottles can be reused:
  - Waste glass can be crushed, melted and reused.
  - Some waste plastic can be recycled to make fleece material.
  - Metals can be recycled by melting them down and then making them into new objects.

Quick Test

1. How does sacrificial protection protect steel objects?
2. Why does aluminum appear to be less reactive than its position in the reactivity series suggests it should be?
3. How can LCAs help people make good decisions about which products to buy?

Key Point

Unscrupulous advertisers may use shortened or abbreviated LCAs, which could lead to misleading claims being made about a particular product.

Key Point

You need to be able to describe and evaluate different methods that can be used to tackle problems caused by human impacts on the environment.

Key Point

Recycling generally uses far less energy than the initial extraction and production processes. As a result, less fossil fuel is burned and less greenhouse gases are released into the atmosphere. It also preserves our reserves of raw materials for the future.

Key Words

corrode
rusting
sacrificial protection
galvanising
composite material
life cycle assessment (LCA)

The Haber Process

You must be able to:

- Recall the conditions used in the Haber process
- HT Explain why these conditions are chosen for the Haber process
- Explain why NPK fertilisers are useful
- Explain how fertilisers are made.

The Haber Process

- **Reversible reactions** may not go to completion.
- However, they can still be used efficiently in continuous processes, such as the **Haber process**.
- The Haber process is used to manufacture ammonia, which is used to produce nitrogen-based fertilisers.
- The raw materials for this process are:
  - purified nitrogen – from the fractional distillation of liquid air
  - hydrogen – from natural gas or steam.
- The nitrogen and hydrogen are passed over an iron **catalyst** at a:
  - moderate temperature (about 450°C)
  - high pressure (about 200 atmospheres).
- Some of the hydrogen and nitrogen reacts to form ammonia.
- Because the reaction is reversible, some of the ammonia produced will break down again into nitrogen and hydrogen.
- On cooling, the ammonia liquefies and can be removed from the mixture.
- The unreacted nitrogen and hydrogen are recycled.

nitrogen + hydrogen $\rightleftharpoons$ ammonia

$$N_2 + 3H_2 \rightleftharpoons 2NH_3$$

Key Point

The Haber process is used to manufacture ammonia.

Key Point

The exact conditions used commercially in the Haber process vary slightly with the availability and cost of raw materials and energy supplies.

HT Choosing the Conditions

- The reaction between nitrogen and hydrogen is **exothermic**.
- A high reaction temperature would give a fast rate of reaction but a low yield of ammonia.
- A low reaction temperature would give a high yield of ammonia but a very slow reaction.
- In practice, a moderate temperature is used.
- A high pressure encourages a high yield of ammonia.
- The iron catalyst increases the rate of the reaction.
- These reaction conditions are chosen to produce a reasonable yield of ammonia quickly. Even so, only some of the hydrogen and nitrogen react together to form ammonia.

NPK Fertilisers

- Fertilisers are used to replace the essential elements in soil that are used up by plants as they grow.
- Plants absorb these chemicals through their roots, so fertilisers must be soluble in water.
- To grow well, plants need nitrogen, N, phosphorus, P, and potassium, K.
- **NPK fertilisers** contain compounds of all three of these elements.
- NPK fertilisers are formulations of various salts that are mixed together to give the appropriate percentage of each element.
- The NPK rating of a fertiliser consists of three numbers, e.g. 16–4–10:
  - the first number gives the percentage of nitrogen
  - the second number relates to the amount of phosphorus
  - the third number relates to the amount of potassium.
- Ammonia is an alkaline gas that dissolves in water.
- It is mainly used in the production of fertilisers, to increase the nitrogen content of the soil.
- Ammonia:
  - can be oxidised to produce nitric acid
  - can neutralise nitric acid to produce ammonium nitrate.

ammonia + nitric acid ⟶ ammonium nitrate

$$NH_3(aq) + HNO_3(aq) \longrightarrow NH_4NO_3(aq)$$

- Ammonium nitrate is a fertiliser rich in nitrogen.
- Nitrogen-based fertilisers are important chemicals because they increase the yield of crops.
- Potassium chloride and potassium sulfate are soluble and can be used directly as fertilisers.
- **Phosphate rock** contains high levels of phosphorus compounds.
- However, these compounds are insoluble, so they cannot be used directly as fertilisers.
- The phosphate rock can be processed to make useful new products:
  - It can be treated with nitric acid to produce phosphoric acid and calcium nitrate. Phosphoric acid is then neutralised with ammonia to produce ammonium phosphate.
  - It can be reacted with sulfuric acid to make single superphosphate – a mixture of calcium phosphate and calcium sulfate (sulfur is also needed to make plants grow well).
  - It can be reacted with phosphoric acid to make triple superphosphate (calcium phosphate).

Key Point

Fertilisers replace the essential elements used up by plants as they grow.

Quick Test

1. Name the catalyst used in the Haber process.
2. In the Haber process, how is it possible to remove the ammonia from the reaction mixture?
3. Why must fertilisers be soluble?
4. What does NPK stand for?
5. How is single superphosphate made?

Key Words

reversible reaction
Haber process
catalyst
 **exothermic**
NPK fertilisers
phosphate rock

Review Questions

Exothermic and Endothermic Reactions

1 Complete the energy profile diagram in **Figure 1** to show an endothermic reaction.
Include the:

- Reactants
- Products
- Activation energy
- Energy change of the reaction.

Figure 1

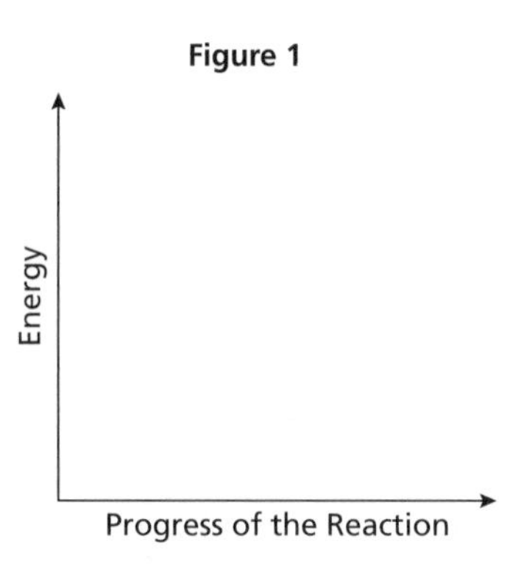

[4]

2 Which of the following types of reaction is an example of an exothermic reaction?
Tick **one** box.

The reaction between citric acid and sodium hydrogen carbonate ☐

The neutralisation of hydrochloric acid by sodium hydroxide solution ☐

Dissolving ammonium nitrate crystals in water ☐

The thermal decomposition of calcium carbonate ☐ [1]

Total Marks .... / 5

Fuel Cells

1 A student carries out an experiment to find out whether changing the metal powder added to a solution of nitric acid affects the temperature change for the reaction.

a) Name the independent variable in this experiment. [1]

b) Why does the student use powdered metals in this experiment? [1]

c) Complete the results in **Table 1** below.

Table 1

Metal	Temperature at the Start (°C)	Temperature at the End (°C)	Temperature Change (°C)
magnesium	21.0	35.5	
zinc	22.0		11.0
iron		29.5	8.5

[3]

Total Marks .... / 5

Rate of Reaction

1 Which of the following changes would decrease the rate of a chemical change?
Tick **one** box.

Increasing the concentration ☐ Increasing the temperature by 20°C ☐

Adding a catalyst ☐ Increasing the size of particles ☐ [1]

2 Explain, in terms of collision theory, why increasing the concentration of the reactants increases the rate of a chemical change. [2]

3 Catalysts are very important in industry.

a) Name the catalyst used in the manufacture of ammonia. [1]

b) Catalysts increase the rate of chemical reactions.

Why do manufacturers want to increase the rate of chemical reactions? [1]

c) Explain how catalysts work. [2]

4 The rate of a chemical reaction can be calculated directly from graphs.

a) Write the equation for the mean rate of reaction. [1]

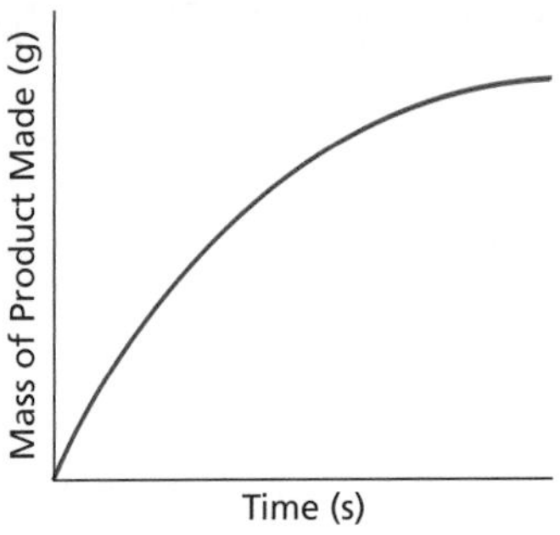

b) The graph in **Figure 1** shows the rate of reaction is fastest at the start of the experiment.

Explain why the rate of reaction is faster at the start of the reaction? [2]

c) HT Explain how the rate of reaction can be calculated from the graph in **Figure 1**. [2]

Figure 1

Mass of Product Made (g)

Time (s)

Total Marks .... / 12

Reversible Reactions

1 **a)** What does the symbol $\rightleftharpoons$ mean? [1]

b) What is happening when equilibrium is achieved in a reaction? [1]

Total Marks .... / 2

Review Questions

Alkanes

1. Which of the substances below is a saturated hydrocarbon?
Tick **one** box.

CH_4 ☐ C_2H_4 ☐ C_3H_6 ☐ C_5H_{10} ☐ [1]

2. Crude oil can be separated into different parts by fractional distillation.

   a) Explain how crude oil is separated by fractional distillation. [3]

   b) Give **one** use for the diesel fraction of crude oil. [1]

3. Propane, C_3H_8, is used as a fuel in camping stoves.

   a) Complete the equation below for the complete combustion of propane.

   $$C_3H_8 + \_\_\_\_ O_2 \rightarrow \_\_\_\_ CO_2 + \_\_\_\_ H_2O$$ [3]

   b) Incomplete combustion of propane can produce carbon monoxide.

   Why is the production of carbon monoxide undesirable? [1]

4. Fractional distillation of crude oil produces short-chain hydrocarbons.

   Which of the properties below do short-chain hydrocarbons have at room temperature?
   Tick **one** box.

   Viscous ☐ Hard to ignite ☐

   High melting points ☐ Gases or volatile liquids ☐ [1]

5. There is a high demand for short-chain alkanes.

   Name the process used to produce short-chain hydrocarbons from long-chain hydrocarbons.
   Tick **one** box.

   Fractional distillation ☐ Addition polymerisation ☐

   Cracking ☐ Condensation polymerisation ☐ [1]

6. Many fuels contain sulfur.
When sulfur is burned a gaseous compound is made.

   This gas reacts with water to form a compound that can damage the environment.

a) Name the gas formed when sulfur is burned. [1]

b) Why does the compound formed when the gas from part **a)** reacts with water cause damage to the environment? [2]

Total Marks .... / 14

Alkenes

1 Alkenes are unsaturated hydrocarbons.

a) Give the general formula for alkenes. [1]

b) What does 'unsaturated' mean? [1]

c) Propene is the second member of the alkene homologous series.

What is the functional group of the alkene homologous series? [1]

d) What is the formula of propene? [1]

2 A student has samples of propene and propane.

Describe the test that the student should carry out to identify the two samples. [4]

3 Name the type of bonds found in alkene molecules.
Tick **one** box.

Metallic and single covalent ☐

Single covalent and double covalent ☐

Ionic and double covalent ☐

Double covalent and metallic ☐ [1]

4 Hydrogen can be added to alkene molecules to form alkane molecules.

a) Name the catalyst used in this reaction. [1]

b) Complete the balanced symbol equation to show what happens when butene reacts with hydrogen.

$$C_4H_8 + H_2 \rightarrow \text{.......}$$ [1]

c) Write a word equation for this reaction. [1]

5 Bromine water can be added to alkene molecules to produce a halogenoalkane.

a) Complete the balanced symbol equation to show what happens when propene reacts with hydrogen.

$$\ldots\ldots + Br_2 \rightarrow C_3H_6Br_2$$ [1]

b) What would happen to the colour of the bromine water during this reaction? [1]

Total Marks .... / 14

Organic Compounds

1 An organic compound has the formula $CH_3CH_2CH_2OH$.

What type of compound is it?
Tick **one** box.

Alkane	☐	Ester	☐
Alcohol	☐	Carboxylic acid	☐

[1]

2 HT Diol and dicarboxylic acid molecules can be reacted together to form a condensation polymer.

a) What is a 'dicarboxylic acid'? [1]

b) What type of organic compound is formed in this reaction? [1]

c) **Figure 1** shows the repeating unit of the compound formed in the reaction.

Circle the functional group. [1]

Figure 1

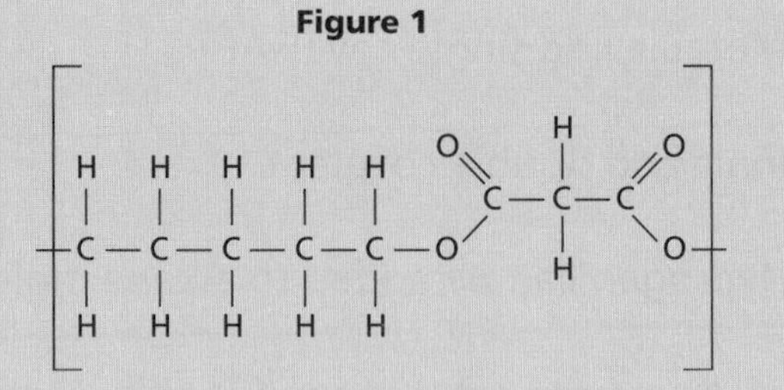

3 HT Amino acids such as glycine and alanine contain two different functional groups.

Identify the two functional groups found in amino acids.
Tick **one** box.

Carboxylic acid and amine	☐	Ester and carboxylic acid	☐
Amine and ester	☐	Alkenes and amines	☐

[1]

4 An organic compound has the formula CH_3CH_2COOH.

a) What is the functional group of this organic compound? [1]

b) Name this organic compound. [1]

c) Suggest the pH of an aqueous solution of this organic compound. [1]

Total Marks .... / 8

Polymerisation

1 Poly(butene) is an addition polymer.

a) Give the formula for the monomer that is used to produce poly(butene). [1]

b) Explain how poly(butene) is formed. [2]

c) What is the atom economy of this reaction? [1]

2 Thermosoftening plastics, such as poly(ethene), are widely used.

a) Explain why poly(ethene) softens when it is heated. [2]

b) Melamine is a thermosetting plastic.

Suggest a possible use for melamine. [1]

c) Explain why melamine does *not* soften when it is heated. [2]

3 Lots of ethene molecules can be joined together to form a polymer.

a) Name the polymer made in this reaction. [1]

b) Complete the equation below to show how this polymer is made.

```
     H   H
     |   |
n    C = C    →
     |   |
     H   H
```

[1]

c) What is the atom economy of this reaction?
You must explain your answer. [2]

Total Marks .... / 13

Practice Questions

Chemical Analysis

1 **a)** Define the term 'pure'. [1]

b) A substance melts at 34°C.

Is this substance a mixture?
You must explain your answer. [2]

2 A student analysed the colourings added to three fizzy drinks.
Figure 1 shows the results.

a) Name the technique that the student used to analyse the colourings. [1]

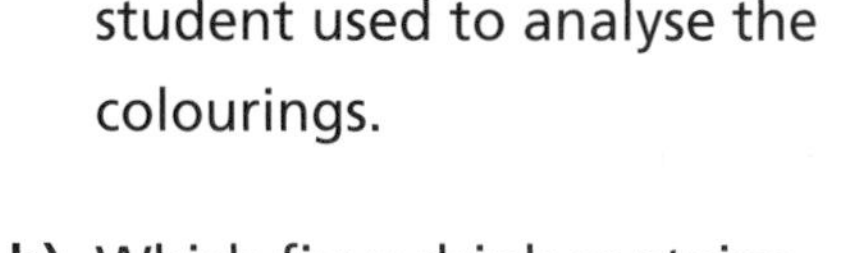

b) Which fizzy drink contains a pure colouring?
Give a reason for your answer. [2]

c) The R_f value can be used to identify the components in the colouring added to fizzy drinks.

Figure 1

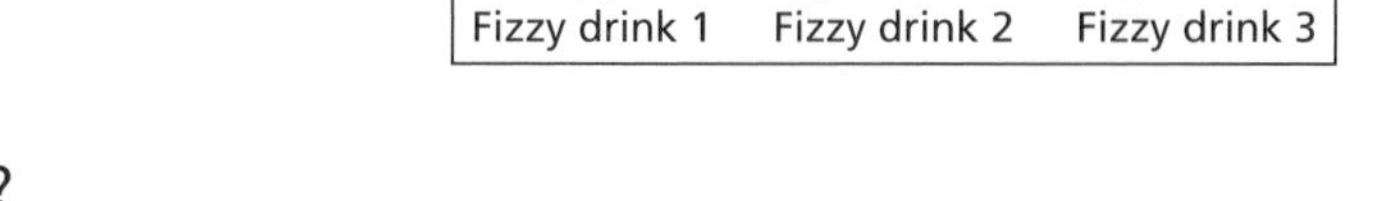

What is the R_f value of component **A**? [2]

d) Which other fizzy drink also contains component **A**?
Explain your answer. [2]

Total Marks .... / 10

Identifying Substances

1 Which of the following flame colours would show that a metal compound contains potassium?
Tick **one** box.

Green ☐ Red ☐ Yellow ☐ Lilac ☐ [1]

2 Sodium hydroxide solution can be used to identify metal ions.

Which of the following precipitate colours would show that a metal compound contained Fe^{2+} ions?
Tick **one** box.

Blue ☐ Green ☐ White ☐ Brown ☐ [1]

3 A sample contains a mixture of sodium chloride and potassium chloride.

a) How could a student prove that the sample contains chloride ions?
Describe the chemicals that the student should add and the results they would expect to see. [3]

b) Explain why a flame test may not be able to identify both the compounds present in this mixture. [3]

4 **a)** Which of the following methods could be used to test for the presence of sulfate ions in a solution?
Tick **one** box.

Add aqueous silver nitrate. ☐

Add dilute hydrochloric acid then barium chloride solution. ☐

Add sodium hydroxide solution. ☐

Carry out a flame test. ☐ [1]

b) Which of the following results would confirm the presence of sulfate ions?
Tick **one** box.

Green flame ☐ Cream precipitate ☐

White precipitate ☐ Yellow precipitate ☐ [1]

5 Flame tests can be used to identify the metal ions present in compounds.

a) To carry out a flame test a nichrome wire is heated in a Bunsen flame and then placed into concentrated hydrochloric acid.

i) Why is a nichrome wire used? [1]

ii) Why is the wire placed in the Bunsen flame? [1]

Practice Questions

b) The wire is then placed into the compound and held in the hottest part of a Bunsen flame.

What colour would be seen if the compound contained copper ions? [1]

6 Describe how a student could confirm that a compound contained carbonate, CO_3^{2-}, ions? [3]

Total Marks .... / 16

The Earth's Atmosphere

1 How many years ago was the Earth formed?
Tick **one** box.

4.6 billion ☐ 4.6 million ☐ 2.7 billion ☐ 200 million ☐ [1]

2 Complete **Table 1** to show the gases present in the Earth's atmosphere today.

Table 1

Gas	Percentage (%)
	78
	21
Noble gases / water vapour / carbon dioxide	1

[2]

3 The atmosphere of the Earth has changed over time.

a) What was the main gas in the Earth's early atmosphere? [1]

b) Which planets have an atmosphere similar to Earth's early atmosphere? [2]

c) How has the evolution of algae and plants affected the level of oxygen in the Earth's atmosphere? [2]

4 During photosynthesis carbon dioxide and water react to produce glucose and oxygen.

a) Write a word equation for this reaction. [1]

b) Balance the equation below to show what happens during photosynthesis.

$$\ldots CO_2 + \ldots H_2O \rightarrow C_6H_{12}O_6 + 6O_2$$ [2]

c) Explain how photosynthesis has changed the level of carbon dioxide in the Earth's atmosphere. [2]

Total Marks .... / 13

Greenhouse Gases

1. Methane is a greenhouse gas. Name **one** other greenhouse gas. [1]

2. Why has the level of methane in the atmosphere been increasing in recent times?
Tick **one** box.

Deforestation	☐	Combustion of fossil fuels	☐
More animal farming	☐	Photosynthesis	☐

[1]

3. **a)** Suggest why it is difficult to produce an accurate model of how increasing levels of carbon dioxide might lead to global climate change. [1]

   b) Why might someone who worked for an oil company give misleading information to the media about the impact of burning fossil fuels on global climate change? [1]

   c) Who would you trust to give valid information about the impact of burning fossil fuels on global climate change? [1]

4. Global climate change could cause sea levels to rise.

   a) Suggest how a rise in sea levels could cause damage. [1]

   b) How could global climate change affect the distribution of wildlife? [2]

   c) What does ‘CCS’ stand for? [1]

   d) How could CCS help prevent global climate change? [1]

5. A cotton bag has a label, which says the bag is carbon-neutral.

   a) What is the ‘carbon footprint’ of a product? [2]

   b) What does the term 'carbon-neutral' mean? [1]

   c) Suggest why it is a good idea that the bag has the carbon-neutral label. [1]

   d) How might carbon taxes reduce the likelihood of global climate change? [2]

6. If the average global temperature increases, this could cause global climate change.

   a) How might this affect the amount of rainfall? [2]

   b) How could this cause a food shortage for people living in some areas? [2]

Total Marks .... / 20

Practice Questions

Earth's Resources

1. Water is essential for life.

   a) What is 'fresh water'? [1]

   b) What is good quality water that is safe to drink called? [1]

   c) Water is collected from sources away from polluted areas and treated to make it safe to drink. In one of the steps, it is passed through filter beds.

   Why is it passed through filter beds? [1]

   d) Why is chlorine added to water? [1]

2. Which of these substances can be added to water to sterilise it?
   Tick **one** box.

Ozone	☐	Salt	☐
Carbon dioxide	☐	Oxygen	☐

[1]

3. Copper is a very useful metal. It is extracted from its ores.

   a) What is an 'ore'? [1]

   b) Copper can be extracted by smelting.

   Describe what happens in the smelting process. [2]

   c) HT Copper can also be extracted by bioleaching.

   Describe how copper is extracted by bioleaching. [2]

Total Marks .... / 10

Using Resources

1. Metals corrode in the environment.

   a) When metals corrode, which substances in the environment do they react with? [2]

   b) Iron objects can be protected from rusting by coating them with a layer of zinc.

   What is the coating process called? [1]

c) Describe what happens if the layer of zinc is scratched. [2]

2. Aluminium is a fairly reactive metal, but it is used to make watering cans and greenhouse frames.

   Why can aluminium be used in these ways? [2]

3. Iron corrodes (or rusts) faster than most other transition metals.
   Iron can be made into the alloy steel.
   Expensive steel objects can be protected by sacrificial protection.

   Describe what happens in sacrificial protection of steel and explain how it works. [3]

Total Marks .... / 10

The Haber Process

1. The Haber process is used to manufacture ammonia: $N_2(g) + 3H_2(g) \rightleftharpoons 2NH_3(g)$
   It is an exothermic reaction.

   a) Write a word equation for this reaction. [1]

   b) The raw materials for the Haber process are nitrogen and hydrogen.
   Hydrogen is obtained from natural gas or steam.

   Where is nitrogen obtained from? [1]

   c) What pressure is used in the Haber process? [1]

   d) HT Why is this pressure used? [2]

   e) How is ammonia removed from the reaction mixture of ammonia, nitrogen and hydrogen? [2]

2. Fertilisers are used by farmers and gardeners.

   a) Why do farmers and gardeners use fertilisers? [1]

   b) Why must fertilisers be soluble? [1]

   c) Which elements are found in NPK fertilisers? [1]

Total Marks .... / 10

Review Questions

Chemical Analysis

1 Chromatography can be used to separate mixtures of coloured substances.

The mixtures are separated because each component in the mixture distributes itself between the stationary phase and the mobile phase differently.

a) **i)** What is the stationary phase in paper chromatography? [1]

ii) What is the mobile phase in paper chromatography? [1]

The R_f value can be used to identify the components in a soluble food colouring. **Figure 1** shows the chromatogram for a sample of food colouring.

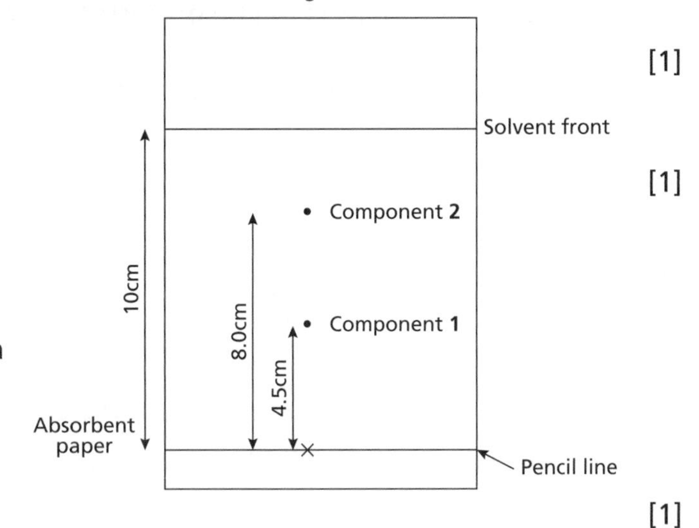

b) Why is the line at the bottom of the absorbent paper drawn in pencil? [1]

c) What is the R_f value for Component 1? [2]

d) How does this chromatogram show that the food colouring tested is a mixture and not a pure substance? [1]

2 Standard laboratory equipment can be used to identify chemicals.
However, modern instrumental methods are often preferred and they produce very accurate results.

a) Give **one** other advantage of modern instrumental methods over traditional methods of analysis. [1]

b) Flame emission spectroscopy is a useful modern instrumental method.
It produces a line spectrum.

What information does the line spectrum provide? [2]

Total Marks .... / 9

Identifying Substances

1. When calcium carbonate is added to hydrochloric acid, a chemical reaction takes place. One of the products is carbon dioxide.

$$CaCO_3(s) + 2HCl(aq) \rightarrow CaCl_2(aq) + H_2O(l) + CO_2(g)$$

a) What does the state symbol (s) indicate? [1]

b) What does the state symbol (g) indicate? [1]

c) Describe how a student would prove that carbon dioxide was produced in this experiment. [2]

2. Which of the following flame colours would show that a metal compound contained sodium? Tick **one** box.

Yellow ☐ Lilac ☐ Green ☐ Red ☐ [1]

3. Sodium chloride and sodium bromide are both white solids.

a) Explain why flame tests cannot be used to identify each of the compounds. [2]

b) Chloride ions and bromide ions can be identified by adding nitric acid and silver nitrate solution to solutions containing the halide ions.

Explain how a student could use the results from this test to identify the two compounds. [2]

4. A solution of sulfuric acid, H_2SO_4, is a strong acid. It contains sulfate, SO_4^{2-}, ions.

a) Name the chemicals that are added to the sulfuric acid solution to test for sulfate ions. [2]

b) Describe and name the compound formed. [2]

5. Sodium hydroxide solution can be used to identify metal ions.

Which of the following precipitate colours would show that a metal compound contained copper(II) ions? Tick **one** box.

Blue ☐ Green ☐ White ☐ Brown ☐ [1]

6. Describe how a student could show that a gas produced in a reaction is oxygen. [2]

7 Chlorine is a toxic gas.
Small amounts of chlorine are added to water to kill microorganisms that could harm people.

Describe how a student could show that a gas produced in a reaction is chlorine. [2]

Total Marks .... / 18

The Earth's Atmosphere

1 **a)** How many years ago did algae evolve and start producing oxygen by photosynthesis?
Tick **one** box.

4.6 billion ☐ 2.7 billion ☐

4.6 million ☐ 200 billion ☐ [1]

b) Complete the word equation to sum up what happens during photosynthesis.

.... + water → glucose + ... [2]

2 The Earth's atmosphere has changed over time.

a) How was the Earth's early atmosphere formed? [1]

b) What was the main gas in the Earth's early atmosphere? [1]

c) The level of nitrogen has increased over time.
Some nitrogen was produced when the gases in the early atmosphere reacted with oxygen.

By what other means was nitrogen produced? [1]

d) How has the formation of sedimentary rocks, like limestone, affected the Earth's atmosphere? [2]

3 Which of these planets has an atmosphere that is mainly carbon dioxide?
Tick **one** box.

Jupiter ☐ Saturn ☐

Mars ☐ Mercury ☐ [1]

4. The Earth's atmosphere has remained almost constant for around 200 million years. However the level of carbon dioxide in the atmosphere has risen sharply in the last 100 years.

   a) How has the use of coal caused an increase in the level of carbon dioxide? [2]

   b) Describe how coal is formed. [3]

5. Some of the carbon dioxide released into the atmosphere reacts with sea water. As a result, there is a smaller increase in carbon dioxide levels than might be expected.

   a) Give **two** types of products that can be formed when carbon dioxide reacts with sea water. [2]

   b) How can algae in sea water reduce the levels of carbon dioxide in the atmosphere? [1]

Total Marks .... / 17

Greenhouse Gases

1. Carbon dioxide is a greenhouse gas.

   a) Write down the chemical formula of carbon dioxide. [1]

   b) Explain how greenhouse gases increase the Earth's temperature. [2]

2. Why has the level of carbon dioxide in the atmosphere been increasing in recent times? Tick **one** box.

   Deforestation ☐

   More animal farming ☐

   Decomposition of rubbish in landfill sites ☐

   Photosynthesis ☐ [1]

3. Carbon offsetting schemes allow companies and individuals to invest in environmental schemes.

   a) Which gas are these schemes designed to compensate for? [1]

   b) Some carbon offsetting schemes involve the planting of trees.

   How would this activity offset carbon? [2]

Total Marks .... / 7

Review Questions

Earth's Resources

1. What can be used to sterilise water?
   Tick **one** box.

Infrared radiation	☐	Carbon dioxide	☐
Ultraviolet radiation	☐	Salt	☐

   [1]

2. Copper is a useful metal.

   a) Why is copper a good metal for making wires? [2]

   b) Why is copper a good metal for making saucepans? [1]

   c) Copper can be extracted from its ore by heating with carbon.

   What property of copper means that it can be extracted using this method? [1]

   d) HT Copper can also be extracted by phytomining.

   Describe how copper is extracted by phytomining. [3]

3. Copper can be extracted by phytomining or bioleaching.
   These two methods of extraction are more environmentally friendly than traditional mining methods.

   Explain why traditional methods of mining can cause environmental problems. [3]

Total Marks .... / 11

Using Resources

1. Rusting is the corrosion of iron or steel objects.

   a) Why does covering iron objects with plastic prevent them from rusting? [2]

   b) What happens to the iron if the plastic coating is damaged and why? [2]

Total Marks .... / 4

The Haber Process

1 The Haber process is used to manufacture ammonia.
It is an exothermic reaction.

a) Balance the equation below to sum up the Haber process.

$$N_2 + \_\_\_ H_2 \rightleftharpoons \_\_\_ NH_3$$ [2]

b) What does the symbol $\rightleftharpoons$ mean? [1]

c) The raw materials for the Haber process are nitrogen and hydrogen.
Nitrogen is obtained from fractional distillation of liquid air.

How is hydrogen obtained? [1]

d) Name the catalyst used in this reaction. [1]

e) Why is a catalyst used in this reaction? [1]

f) HT A temperature of 450°C is used in the Haber process.

i) What would happen to the rate of reaction and the yield of ammonia if the temperature was increased? [2]

ii) What would happen to the rate of reaction and the yield of ammonia if the temperature was decreased? [2]

iii) Why is a temperature of 450°C used in the Haber process? [2]

2 Fertilisers have an NPK rating.
Figure 1 shows the NPK rating for three different fertilisers.

Figure 1

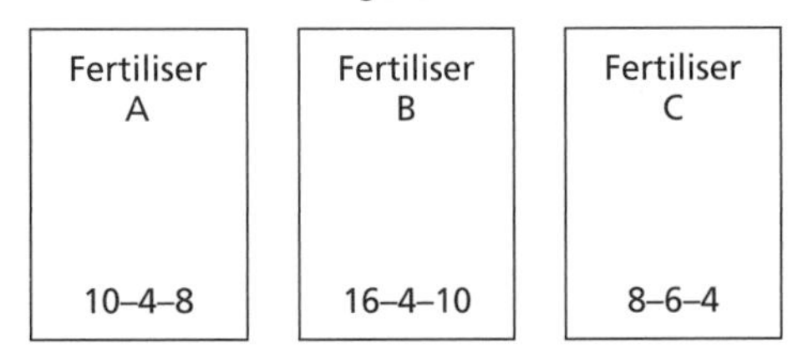

Which fertiliser contains the most potassium? [1]

Total Marks .... / 13

Mixed Exam-Style Questions

1 **Figure 1** shows a potassium atom.

Figure 1

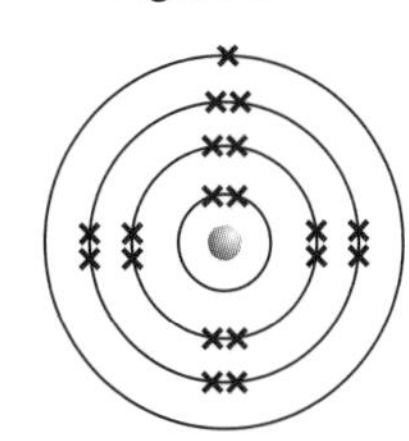

a) Write down the electron configuration of this potassium atom. [1]

b) Explain why potassium belongs to Group 1 of the periodic table. [2]

c) Sodium has an atomic number of 11.

Draw a diagram to show how the electrons are arranged in a sodium atom. [1]

d) Explain why the chemical reactions of potassium and sodium are similar. [1]

2 Transition metals and their compounds are often good catalysts.

What is a catalyst? [2]

3 What is an 'ion'? [2]

4 An ion of lithium has the symbol ${}^{7}_{3}Li^{+}$.

Give the number of protons, electrons and neutrons in this ion of lithium. [3]

5 Hydrogen fuel cells are used to produce electricity.

a) Why are hydrogen fuel cells considered to be non-polluting? [1]

b) HT In a hydrogen fuel cell, the fuel is oxidised electrochemically to produce a potential difference.

At the anode: $2H_2 \rightarrow 4H^+ + 4e^-$
At the cathode: $O_2 + 4H^+ + 4e^- \rightarrow 2H_2O$

Write the overall equation for the reaction that takes place in a hydrogen fuel cell. [2]

6 Tin is extracted from its ore, tin oxide, by heating the tin oxide with carbon.

a) Complete the word equation for the extraction of tin.

tin oxide + carbon → .... + ... [2]

b) In terms of oxidation and reduction, explain what happens to tin oxide and carbon in this reaction. [2]

7 Which salt is produced when sulfuric acid is neutralised by sodium hydroxide?
Tick **one** box.

Sulfur sulfate ☐ Sodium sulfate ☐

Sodium chloride ☐ Sulfur chloride ☐ [1]

8 Graphene has some special properties, which could make it very useful in the future.

Tick each of the properties of graphene.

Property	Does graphene have this property?
Very strong	
Low melting point	
Good electrical conductor	
Good thermal conductor	
Opaque	
Nearly transparent	
Excellent electrical insulator	
Fragile	

[4]

9 Potassium fluoride is an ionic compound.

a) Suggest why potassium fluoride has a high melting point. [2]

b) Does potassium fluoride conduct electricity when solid?
You must give a reason for your answer. [1]

c) Does potassium fluoride conduct electricity when it is molten?
You must give a reason for your answer. [1]

d) Does potassium fluoride conduct electricity when dissolved in water to form an aqueous solution?
You must give a reason for your answer. [1]

Mixed Exam-Style Questions

10 A calcium atom can be represented by:

$^{41}_{20}Ca$

a) How many protons does this atom of calcium have? [1]

b) How many neutrons does this atom of calcium have? [1]

c) How many electrons does this atom of calcium have? [1]

11 Ammonia, NH_3, is produced from nitrogen and hydrogen.

Relative atomic masses (A_r): N = 14, H = 1

a) Calculate the relative molecular mass of ammonia, NH_3. [2]

b) HT Calculate the mass of 1.00 mole of ammonia, NH_3. [1]

c) HT Calculate the mass of 0.5 moles of ammonia, NH_3. [1]

12 Balance the following symbol equations.

a) .... $Ca + O_2 \rightarrow$ .... CaO [2]

b) .... $Na + Br_2 \rightarrow$ .... $NaBr$ [2]

c) $H_2 + Br_2 \rightarrow$ .... HBr [1]

d) .... $H_2 + N_2 \rightarrow$ .... NH_3 [2]

e) .... $K + I_2 \rightarrow$ .... KI [2]

13 What technique could be used to extract a sample of pure water from a solution of salt and water?
Tick **one** box.

Crystallisation	☐	Filtering	☐
Distillation	☐	Chromatography	☐

[1]

14 **Figure 2** shows a lithium atom.

Figure 2

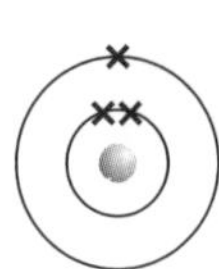

a) Write down the electron configuration of this lithium atom. [1]

b) Explain why lithium belongs to Group 1 of the periodic table. [1]

15 Everything is made of matter. There are three states of matter.

a) Complete **Figure 3** to show how the particles are arranged in solids, liquids and gases. Use the symbol 'o' to represent a particle.

Figure 3

Solid	Liquid	Gas

[3]

b) Explain how the particles are moving in a gas. [2]

c) Complete the sentence below.

When substances boil, they change from the .... to the .... state. [2]

d) Why does water have a much lower boiling point than iron?

Your answer should include:

- The type of bonding and the structure of each substance.
- The strength of these forces of attraction.

[6]

16 **a)** HT In terms of electron transfer, explain why bromide ions have a 1– charge. [2]

b) HT In terms of electron transfer, explain why calcium ions have a 2+ charge. [2]

17 **a)** What is the chemical symbol for silver? [1]

b) Name the type of bonding in silver. [1]

c) Pure silver is too soft for many uses.

Why is pure silver soft? [2]

d) Silver is sometimes made into an alloy.

What is an alloy? [1]

Mixed Exam-Style Questions

18 Balanced symbol equations can be used to sum up what happens in chemical reactions.

Balance these symbol equations:

a) $Mg + \ldots\ldots HCl \rightarrow MgCl_2 + H_2$ [1]

b) $\ldots\ldots HgO \rightarrow \ldots\ldots Hg + O_2$ [2]

19 Poly(pentene) is an addition polymer.

a) Give the formula of the monomer that is used to produce poly(pentene). [1]

b) Explain how poly(pentene) is produced. [3]

c) What is the atom economy of this reaction? [1]

20 Name the type of bond found in alkane molecules.
Tick **one** box.

Double covalent ☐ Single covalent ☐

Giant ionic ☐ Giant metallic ☐ [1]

21 **Figure 4** shows the outer electrons of a magnesium atom and an oxygen atom.

Figure 4

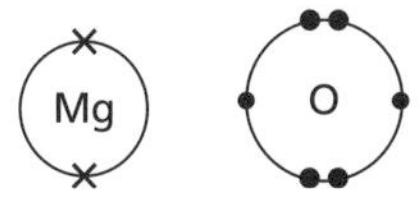

Mg O

a) **i)** Which group of the periodic table does magnesium belong to? [1]

ii) Which group of the periodic table does oxygen belong to? [1]

iii) Draw a diagram to show the electronic structure of a magnesium ion and an oxide ion. Show the outer shell of electrons only. [2]

b) The compound magnesium oxide is a solid at room temperature.

What type of structure does magnesium oxide have?
Tick **one** box.

Metallic ☐ Giant covalent ☐

Simple molecular ☐ Giant ionic ☐ [1]

c) Explain why magnesium oxide does not conduct electricity when solid but does when molten. [2]

22 Ethane, C_2H_6, is a simple molecule.

a) What sort of bonding is found in ethane molecules? [1]

b) **Figure 5** shows the outer electrons of a carbon atom and a hydrogen atom.

Figure 5

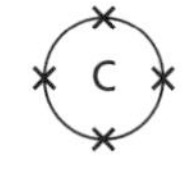

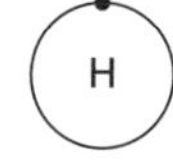

Complete **Figure 6** to show the electron arrangement in an ethane molecule.

Figure 6

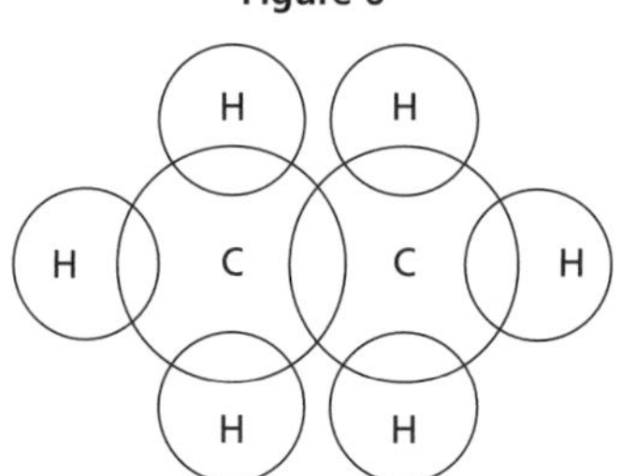

[1]

23 Ethane, C_2H_6, is a fuel.

a) By what process is ethane obtained from crude oil? [1]

b) Complete the equation below for the complete combustion of ethane.

$$2C_2H_6 + \ldots\ldots O_2 \rightarrow \ldots\ldots CO_2 + \ldots\ldots H_2O$$ [3]

c) Incomplete combustion of ethane can produce carbon monoxide.

Why is the production of carbon monoxide a concern? [1]

24 HT Titration can be used to measure how much alkali is needed to neutralise an acid.
25.0cm^3 of sodium hydroxide was placed in a flask.
The sodium hydroxide had a concentration of 0.15mol/dm^3.
This required 28.0cm^3 of nitric acid solution for complete neutralisation.
The equation for the reaction can be summed up by the equation:

$$HNO_3 + NaOH \rightarrow NaNO_3 + H_2O$$

a) How many moles of sodium hydroxide were used in this reaction? [2]

b) How many moles of nitric acid were used in this reaction? [1]

c) What was the concentration of the nitric acid? [2]

25 Alkenes are unsaturated hydrocarbons.

a) Give the general formula of alkenes. [1]

b) What is the functional group of the alkene homologous series? [1]

c) What is the formula of butene? [1]

26 A student calculated that their experiment should produce 15.0g of product. However, after carrying out the reaction, only 8.5g of product was actually produced.

Calculate the percentage yield of this reaction. [2]

27 HT The Avogadro constant has a value of 6.02×10^{23}.

a) How many atoms are present in 39g of potassium? [1]

b) How many atoms are present in 16g of sulfur? [1]

28 Which of the substances below is a saturated hydrocarbon?
Tick **one** box.

C_3H_6 ☐ C_2H_4 ☐ C_3H_8 ☐ C_4H_8 ☐ [1]

29 Lead bromide is an ionic compound. It can be separated by electrolysis.

a) i) Name the element formed at the positive electrode during the electrolysis of molten lead bromide. [1]

ii) Name the element formed at the negative electrode during the electrolysis of molten lead bromide. [1]

b) HT Complete the half equations to show the reactions that take place at each electrode.

i) At the anode: $2Br^- \rightarrow$ .... $+ 2e^-$ [1]

ii) At the cathode: $Pb^{2+} +$ .... $\rightarrow Pb$ [1]

30 HT Diol and dicarboxylic acid molecules can be reacted together to form a polymer.

a) What is a dicarboxylic acid? [1]

b) What type of organic compound is formed in this reaction? [1]

c) What type of reaction occurs to produce this polymer? [2]

d) Complete the equation in **Figure 7** to show a repeat unit for the polymer produced from the diol and dicarboxylic acid.

Figure 7

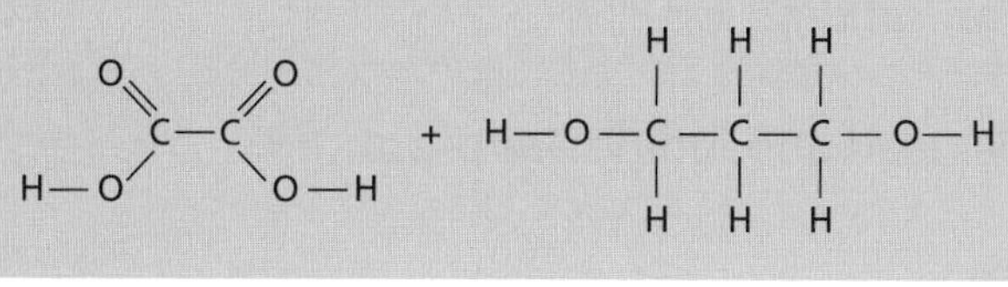

```
   O       O              H   H   H
    \\    //              |   |   |
     C—C        +  H—O—C—C—C—O—H
    /    \                |   |   |
H—O       O—H             H   H   H
```

[1]

31 Gardeners use fertilisers to help their plants.

a) Why do farmers and gardeners use fertilisers? [1]

b) Why should fertilisers be soluble in water? [1]

c) NPK fertilisers are very popular.

What does the K stand for? [1]

d) Fertilisers have an NPK rating.
Figure 8 shows the NPK rating for three different fertilisers.

Figure 8

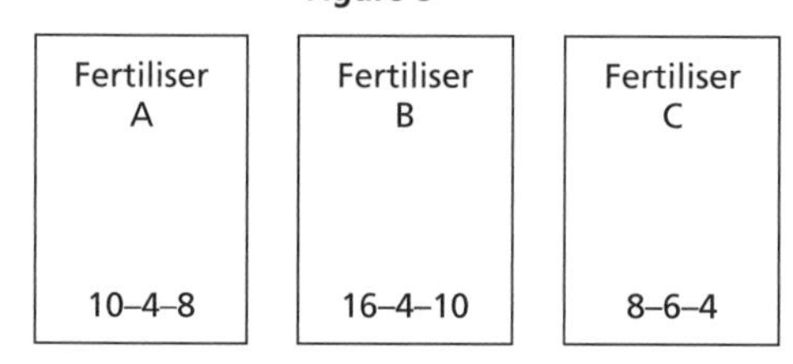

Which fertiliser contains the least nitrogen? [1]

32 An organic compound has the formula $CH_3CH_2CH_2COOH$.

a) What is the functional group of this organic compound? [1]

b) Name this organic compound. [1]

c) Suggest the pH of an aqueous solution of this organic compound. [1]

d) Which of these ions is present in excess in solutions of this compound?
Tick **one** box.

H^+ ☐ H^- ☐ OH^+ ☐ OH^- ☐ [1]

33 Collision theory can be used to explain the rate of a chemical reaction.

a) What **two** things must happen for a chemical reaction to take place? [2]

b) Explain, in terms of collision theory, why increasing the pressure of a reaction that involves gases increases the rate of a chemical reaction. [2]

34 During the electrolysis of molten lead iodide, lead and iodine are produced.

At the anode: $2I^- \rightarrow I_2 + 2e^-$
At the cathode: $Pb^{2+} + 2e^- \rightarrow Pb$

a) Describe what happens to the lead ions during the electrolysis of lead iodide. [2]

b) During the electrolysis of lead iodide, oxidation and reduction take place.

In terms of oxidation and reduction, explain what happens to the iodide ions. [2]

c) Why does the lead iodide have to be molten? [1]

35 Iron can be made into the alloy steel.
Iron and steel objects rust.

a) Expensive steel and iron objects can be protected by sacrificial protection.

Explain how magnesium can be used to protect an expensive steel object by sacrificial protection. [3]

b) Explain why copper is not used to protect an expensive steel object by sacrificial protection. [2]

36 Sodium hydroxide solution can be used to identify metal ions.

Which of the following precipitate colours would show that a metal compound contained Fe^{3+} ions?
Tick **one** box.

Blue ☐ Green ☐ White ☐ Brown ☐ [1]

37 **a)** What is galvanized steel? [1]

b) Why is steel galvanized? [2]

38 Metals such as lead and copper are extracted from their ores.

a) What is an ore? [1]

b) Lead can be extracted by smelting.

Describe what happens during smelting. [2]

c) Why can lead be extracted by smelting? [1]

d) HT Copper can be extracted from its ore by phytomining.

What is phytomining? [1]

e) Aluminium is more reactive than iron, but it is used to make drinks cans and saucepans.

Why can aluminium be used in these ways? [2]

39 Which of the following flame colours would show that a metal compound contained lithium?
Tick **one** box.

Yellow ☐ Lilac ☐ Green ☐ Red ☐ [1]

40 The Haber process is used in industry. It is an exothermic reaction.

$$N_2(g) + 3H_2(g) \rightleftharpoons 2NH_3(g)$$

a) Name the product of this reaction. [1]

b) What does the symbol $\rightleftharpoons$ mean? [1]

c) What is the temperature used in the Haber process? [1]

d) HT How does using a higher temperature affect the rate of reaction and the yield of the product? [2]

e) What happens to unreacted nitrogen and hydrogen? [1]

Mixed Exam-Style Questions

41 Which of the following changes would increase the rate of a chemical change?
Tick **one** box.

Carrying out the reaction in the dark ☐

Reducing the temperature by 10°C ☐

Increasing the surface area of solids ☐

Decreasing the concentration ☐ [1]

42 The equation for the combustion of methane is shown below:

$$CH_4(g) + 2O_2(g) \rightarrow CO_2(g) + 2H_2O(l)$$

Methane is a fuel. It is obtained from crude oil.

a) What is a fuel? [1]

b) By what process is methane obtained from crude oil? [1]

c) Complete the energy profile diagram in **Figure 9** for the burning of methane to show:

- The reactants [1]
- The products [1]
- The activation energy [1]
- The energy change of the reaction. [1]

Figure 9

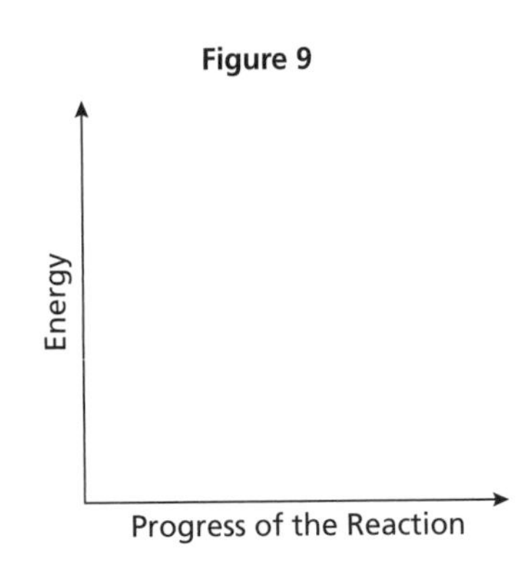

43 Chromatography can be used to separate mixtures of coloured substances.
The R_f value can be used to identify the components.
Figure 10 shows a chromatogram of a colouring ingredient used in fruit-flavoured drinks.

Figure 10

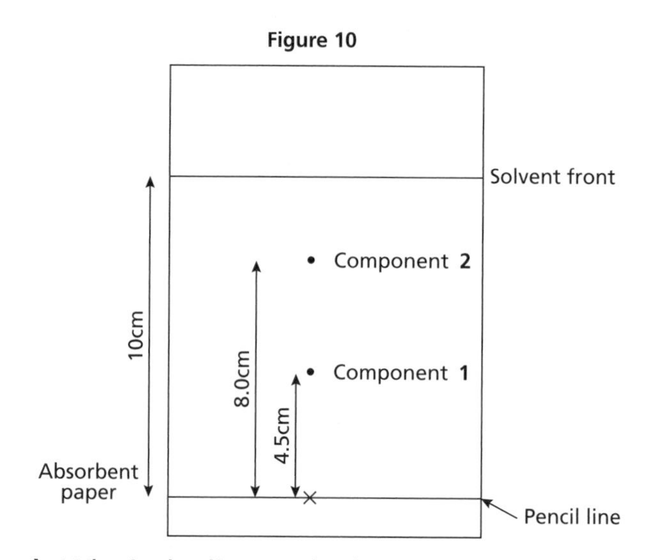

a) Why is the line at the bottom of the absorbent paper drawn in pencil? [1]

b) What is the R_f value for **Component 2?** [2]

c) Explain how this chromatogram shows that the colouring ingredient is a mixture. [1]

44 A solution of sodium sulfate contains sodium, Na^+, and sulfate, SO_4^{2-}, ions.

a) Describe how a student could show that the sodium sulfate solution contains sodium ions. [2]

b) i) Name the chemicals that should be added to the sodium sulfate solution to test for sulfate ions. [2]

ii) Describe and name the compound that will be formed. [2]

45 When magnesium carbonate is added to a beaker of hydrochloric acid a chemical reaction takes place.

$$MgCO_3(s) + 2HCl(aq) \rightarrow MgCl_2(aq) + H_2O(l) + CO_2(g)$$

a) What does the state symbol (s) indicate? [1]

b) What does the state symbol (g) indicate? [1]

c) HT Hydrochloric acid is a strong acid.

What is a strong acid? [2]

d) During the reaction, the mass of the reaction mixture and beaker goes down.

Why does the mass go down? [2]

Total Marks .... / 177

Answers

Pages 6–7 Review Questions

1. a) i) Solid [1]
   ii) liquid [1]
   b) Oxygen **[1]**; carbon dioxide **[1]**
   c) Fuels are substances that can be burned to release energy. [1]
   d) Carbon monoxide [1]
   e) S [1]
   f) Sulfur dioxide [1]
2. a) 3 [1]
   b) Potassium [1]
   c) 5 [1]
3. a) A = alkaline **[1]**; B = alkaline **[1]**; C = neutral **[1]**; D = acidic **[1]**

   Remember, acids have a low pH.

   b) To clean it / stop contamination [1]
   c) Lime / alkali **[1]** (Accept a named alkali)
4. a) Crystallisation / by evaporating the water [1]
   b) Filtration [1]
   c) Three correctly drawn lines **[2]** (1 mark for one correct line)
   Mixture – Contains two or more elements or compounds, which are not chemically joined.
   Compound – Contains atoms of two or more elements, which are chemically joined.
   Element – Contains only one type of atom.
5. a) To mix it / make it react faster [1]
   b) Wear goggles / do not touch the copper sulfate solution [1]
   c) Colour change / temperature change [1]
   d) Copper [1]
   e) iron + copper sulfate → iron sulfate **[1]**; + copper **[1]**

Pages 8–23 Revise Questions

Page 9 Quick Test

1. The smallest part of an element that can exist
2. Elements contain just one type of atom; compounds contain atoms of at least two different elements, which have been chemically combined
3. It contains 1 calcium atom to 1 carbon atom to 3 oxygen atoms
4. Simple distillation

Page 11 Quick Test

1. An atom consisting of tiny negative electrons (the 'plums') surrounded by a sea of positive charge (the 'pudding')
2. 19 protons, 19 electrons and 20 neutrons
3. 19 protons, 18 electrons and 20 neutrons

Page 13 Quick Test

1. Mendeleev
2. They have a full outer shell of electrons, which is a very stable arrangement
3. To stop them reacting with oxygen or water / moisture in the air
4. Form coloured compounds; have ions with different charges; can be used as catalysts

Page 15 Quick Test

1. Liquid
2. Aqueous / dissolved in water
3. Quickly and randomly in all directions
4. To melt, the strong bonds between the ions must be broken. This requires lots of energy, so it only happens at high temperatures.
5. The particle model does not take into account: the forces between the particles; the volume of the particles; or the space between particles.

Page 17 Quick Test

1. Atoms that have gained or lost electrons and now have an overall charge
2. They lose electrons
3. KCl
4. The ions in molten ionic compounds can move about, carrying their charge

Page 19 Quick Test

1. An attraction between the positive metal ions and the delocalised electrons
2. It is an unreactive metal, so it does not react with water, and it can be easily shaped
3. A mixture that contains a metal and at least one other element
4. They are usually stronger and harder than pure metals
5. Hard; resistant to corrosion

Page 21 Quick Test

1. A shared pair of electrons between atoms
2. It is a simple molecule – although there is a strong covalent bond within the molecule, there are only weak intermolecular forces between molecules
3. The larger the molecule, the stronger the intermolecular forces between molecules
4. Each carbon atom forms four strong covalent bonds with other carbon atoms
5. The delocalised electrons can move and carry their charge

Page 23 Quick Test

1. It is a single layer of graphite, just one atom thick
2. Buckminsterfullerene
3. 1–100nm / a few hundred atoms in size
4. They could potentially damage human cells or the environment

Pages 24–29 Practice Questions

Page 24 Atoms, Elements, Compounds and Mixtures

1. Reactants [1]
2. copper + oxygen → copper oxide [1]
3. Atoms are not made or destroyed during reactions [1]
4. Mixtures consist of two or more elements or compounds that are not chemically joined [1]
5. Chromatography [1]
6. Crystallisation [1]
7. They have different boiling points [1]

Page 24 Atoms and the Periodic Table

1.

Subatomic Particle	Relative Mass
proton	1 **[1]**
neutron **[1]**	1
electron	very small **[1]**

2. 19 protons **[1]**; 18 electrons **[1]**; 20 neutrons **[1]**

   A positive (+) ion is formed by the loss of an electron from an atom.

3. 1×10^{-10}m [1]
4. a) 12 [1]
   b) 13 [1]
   c) 12 [1]
5. Isotopes have the same atomic number / number of protons **[1]**; but a different mass number / number of neutrons **[1]**

Page 25 The Periodic Table

1. a) Group 1 [1]
   b) 1 [1]
   c) The outer electron gets further away from the influence of the nucleus **[1]**; so it can be lost more easily **[1]**
   d) Because they react vigorously with oxygen **[1]**; and water **[1]**

Page 26 States of Matter

1. a) Gas [1]
   b) Aqueous / dissolved in water [1]
2. It does not take into account the: forces between the particles **[1]**; that particles, although small, do have some volume **[1]**; the space between particles **[1]**
3. a) Solid [1]
   b) Aqueous / dissolved in water [1]
   c) Gas [1]

Page 26 Ionic Compounds

1. a) It contains lots of strong ionic bonds **[1]**; so lots of energy is needed to overcome the bonds **[1]**
   b) No, the ions cannot move in a solid [1]

c) Yes, the ions can move in a molten state **[1]**

d) Yes, the ions can move in an aqueous solution **[1]**

2. a) An atom that has gained or lost electrons **[1]**
   b) It has gained **[1]**; 1 electron **[1]**
   c) It has lost **[1]**; 2 electrons **[1]**
   d) i) KCl **[1]**
      ii) MgS **[1]**
      iii) CaO **[1]**

> Remember that compounds have no overall charge.

   e) It has lots of strong ionic bonds **[1]**; so lots of energy is required to overcome these bonds **[1]**

Page 27 Metals

1. a) Au **[1]**

> Use your periodic table if you need to.

   b) Metallic **[1]**
   c) Atoms form layers **[1]**; that can slip over each other **[1]**
   d) A mixture that contains at least one metal **[1]**
   e) It has delocalised / free electrons **[1]**; that can move **[1]**

Page 28 Covalent Compounds

1. a) Carbon **[1]**
   b) It has lots of strong covalent bonds **[1]**
   c) It contains no charged particles **[1]**
2. a) It contains no charged particles **[1]**
   b) There are only weak forces of attraction between methane molecules **[1]**; that are easily overcome **[1]**
3. a) i) Group 2 **[1]**
      ii) Group 6 **[1]**
      iii) Correctly drawn outer electrons of a magnesium ion **[1]**; and sulfide ion **[1]**

   $[Mg]^{2+}$ $[S]^{2-}$

   b) Giant ionic **[1]**
   c) The ions **[1]**; cannot move in a solid **[1]**
4. a) A shared pair of electrons **[1]**
   b) Correctly drawn diagram **[1]**

   H N H H

   c) They contain no charged particles **[1]**

Page 29 Special Materials

1. a) 1 to 100nm **[1]**
   b) They could get into and damage human cells / damage the environment **[1]**

Pages 30–43 Revise Questions

Page 31 Quick Test

1. The total mass of the products of a chemical reaction is always equal to the total mass of the reactants.
2. Because no atoms are lost or made, the products of a chemical reaction are made from exactly the same atoms as the reactants.
3. $(1 \times 2) + 16 = 18$
4. $40 + 12 + (16 \times 3) = 100$
5. Because the iron combines with oxygen in the air and the oxygen has mass

Page 33 Quick Test

1. a) $\frac{69}{23} = 3$ moles
   b) $\frac{3}{2} = 1.5$ moles
   c) $1.5 \times (35.5 \times 2) = 106.5$g

Page 35 Quick Test

1. $\frac{1.50}{1.00} = 1.50\text{mol/dm}^3$
2. $\frac{2.00}{2.00} = 1.00\text{g/dm}^3$
3. $\frac{0.20}{0.50} = 0.40\text{mol/dm}^3$

Page 37 Quick Test

1. If the reaction is reversible and it does not go to completion; some product is lost when it is separated from the reaction mixture; some of the reactants react in different ways to the expected reaction
2. percentage yield $= \frac{8.5}{13} \times 100 = 65.4\%$
3. They produce less waste material that could end up damaging the environment
4. 100% (there is only one product)

Page 39 Quick Test

1. Calcium
2. iron + copper sulfate → iron sulfate + copper (the products can be in any order)
3. Lead is extracted from lead oxide by heating it with carbon
4. a) Calcium, Ca, is oxidised – it loses electrons
   b) Hydrogen, H, is reduced – it gains electrons

Page 41 Quick Test

1. pH 7
2. acid + alkali → salt + water (the products can be given in any order)
3. A strong acid is an acid that is fully ionised in water
4. The concentration increases by a factor of 10 × 10, i.e. $10^2 = 100$

Page 43 Quick Test

1. Hydrogen and bromine
2. At the cathode: $2H^+ + 2e^- \rightarrow H_2$
   At the anode: $2Br^- \rightarrow Br_2 + 2e^-$

Pages 44–49 Review Questions

Page 44 Atoms, Elements, Compounds and Mixtures

1. Products **[1]**
2. magnesium + oxygen → magnesium oxide **[1]**
3. Compounds contain atoms of two or more elements that are chemically combined in fixed proportions **[1]**
4. Chromatography **[1]**
5. Atoms are not made or destroyed in chemical reactions / mass is conserved **[1]**
6. a) $H_2 + Br_2 \rightarrow 2HBr$ **[1]**
   b) $2SO_2 + O_2$ **[1]**; $\rightarrow 2SO_3$ **[1]**
   c) $CH_4 + 2O_2$ **[1]**; $\rightarrow CO_2 + 2H_2O$ **[1]**
   d) $N_2 + 3H_2$ **[1]**; $\rightarrow 2NH_3$ **[1]**
   e) $2K + Br_2$ **[1]**; $\rightarrow 2KBr$ **[1]**

Page 44 Atoms and the Periodic Table

1.

Subatomic Particle	Relative Charge
proton	+1 **[1]**
neutron **[1]**	none
electron	–1 **[1]**

2. a) 13 **[1]**
   b) 14 **[1]**
   c) 13 **[1]**
3. 0.1nm **[1]**
4. a) 8 **[1]**
   b) 8 **[1]**
   c) 10 **[1]**

> An oxide ion has gained two electrons.

Page 45 The Periodic Table

1. Good catalysts **[1]**; form ions with different charges **[1]**; form coloured compounds **[1]**
2. a) Group 7 / halogens **[1]**
   b) 17 **[1]**
   c) Isotopes have the same atomic number but a different mass number **[1]**
   d) Same number of protons / 17 protons **[1]**; same number of electrons / 17 electrons **[1]**; different number of neutrons / 18 and 20 neutrons **[1]**

Page 46 States of Matter

1. a) Solid **[1]**
   b) Liquid **[1]**

Answers

Page 46 Ionic Compounds

1. a) An atom that has gained **[1]**; or lost electrons **[1]**
   b) It has lost **[1]**; one electron **[1]**
   c) It has gained **[1]**; two electrons **[1]**
   d) i) $SrCl_2$ **[1]**
      ii) KBr **[1]**
      iii) MgS **[1]**
2. a) i) Group 1 / alkali metals **[1]**
      ii) Group 7 / halogens **[1]**
   b) Correct dot and cross diagrams **[1]**; correct charges **[1]**

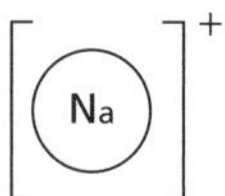

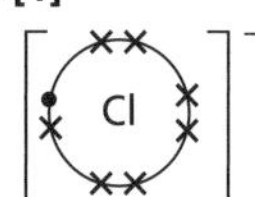

   c) i) Ionic **[1]**
      ii) Lots of strong bonds **[1]**; lots of energy is needed to overcome them **[1]**

Page 47 Metals

1. a) i) Solid **[1]**
      ii) Gas **[1]**
   b) Metallic **[1]**
   c) It has delocalised / free electrons **[1]**; that can move **[1]**
   d) Covalent **[1]**
   e) Only weak forces of attraction between the chlorine molecules **[1]**
   f) Ionic **[1]**
   g) When solid the ions cannot move **[1]**; when molten the ions can move **[1]**
2. a) Metallic **[1]**
   b) Delocalised / free electrons **[1]**; which can move **[1]**

Page 48 Covalent Compounds

1. a) Diamond / graphene / nanotubes **[1]**
   b) 3 **[1]**
   c) It has delocalised / free electrons **[1]**; that can move **[1]**
   d) It has lots of strong covalent bonds **[1]**; and lots of energy is required to overcome these forces of attraction **[1]**
2. a) Group 5 **[1]**
   b) There are only weak forces of attraction between particles **[1]**; which are easily overcome, so little energy is required **[1]**
   c) It does not contain any charged particles **[1]**
3. a) i) 2 **[1]**
      ii) 4 **[1]**
   b) It contains lots of strong covalent bonds **[1]**; so lots of energy is needed to break the bonds **[1]**
4. a) Covalent **[1]**
   b) Correctly drawn diagram **[1]**

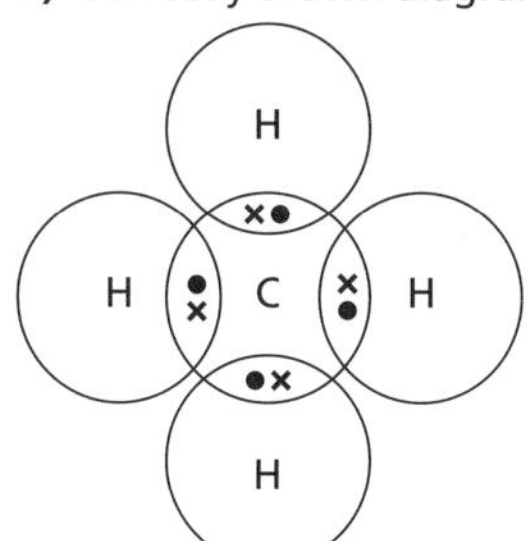

   c) Butane **[1]**; it is a larger molecule **[1]**; so there are stronger forces of attraction between butane molecules **[1]**

Page 49 Special Materials

1. a) Carbon **[1]**
   b) A single layer / one atom thick of graphite **[1]**; carbons are in a hexagonal structure / honeycomb structure **[1]**

Pages 50–55 Practice Questions

Page 50 Conservation of Mass

1. a) 32 + (2 × 16) **[1]**; = 64 **[1]**
   b) 64g **[1]**
   c) 32g **[1]**
2. a) $2e^-$ **[1]**
   b) $2e^-$ **[1]**; Cu **[1]**
3. 164 **[1]**
4. Hydrogen / a gas is made **[1]**; and it escapes from the flask **[1]**

Page 51 Amount of Substance

1. Moles **[1]**
2. a) 6.02×10^{23} **[1]**
   b) 1.204×10^{24} **[1]**

Remember, one mole of any substance contains 6×10^{23} atoms.

3. a) $\frac{19}{38}$ **[1]**; 0.5mol **[1]**
   b) $\frac{22}{44}$ **[1]**; 0.5mol **[1]**
   c) $\frac{17}{17}$ **[1]**; 1.0mol **[1]**
4. a) $\frac{1.8}{12}$ **[1]**; = 0.15mol **[1]**
   b) 44 **[1]**; 0.15 × 44 = 6.6g **[1]**
5. a) There is more than enough / some is left over **[1]**
   b) $\frac{1.8}{18}$ **[1]**; = 0.1mol **[1]**
   c) 0.1mol (same as number of moles of water vapour produced) **[1]**
   d) 0.1 × 2 **[1]**; = 0.2g **[1]**
6. a) 0.6 × 24 **[1]**; = $14.4dm^3$ **[1]**
   b) $\frac{6}{24}$ **[1]**; = 0.25mol **[1]**
7. a) The total mass of the reactants is equal to the total mass of the products / no mass is gained or lost **[1]**
   b) Gas **[1]**
   c) It has reacted with oxygen **[1]**; oxygen has mass **[1]**
   d) 2.0 – 1.2 = 0.8g **[1]**
8. a) 63.5 + ((16 + 1) × 2) **[1]**; = 97.5 **[1]**
   b) 97.5g **[1]**

One mole of a substance is the formula mass in grams.

Page 52 Titration

1. $0.1mol/dm^3$ **[2]** (**1** mark for correct value; **1** mark for correct unit)
2. a) $\frac{20}{1000}$ × 0.2 **[1]**; 0.02 × 0.2 = 0.004mol **[1]**
   b) 0.004mol **[1]**
   c) $\frac{0.004 \times 1000}{18}$ **[1]**; = $0.22mol/dm^3$ **[1]**

Page 53 Percentage Yield and Atom Economy

1. a) The reaction is reversible and does not go to completion. **[1]**
   b) $\frac{9.5}{12}$ **[1]**; = 79% **[1]**

Page 54 Reactivity of Metals

1. a) calcium + oxygen  calcium oxide **[1]**
   b) Calcium is oxidised **[1]**; it loses electrons **[1]**; oxygen is reduced **[1]**; it gains electrons **[1]**

Page 54 The pH Scale and Salts

1. 14 **[1]**
2. H^+ **[1]**
3. a) Concentration **[1]**; of H^+ ions **[1]**
   b) potassium sulfate **[1]**; water **[1]**

Page 55 Electrolysis

1. a) i) Chlorine **[1]**
      ii) Copper **[1]**
   b) i) Cl_2 **[1]**
      ii) $2e^-$ **[1]**

Make sure the equation balances for the species and for the charges.

2. a) Na^+ **[1]**; H^+ **[1]**
   b) Hydrogen **[1]**; sodium is more reactive than hydrogen **[1]**
   c) Chlorine **[1]**; a halide ion is present so a halogen is made **[1]**
3. a) i) They are reduced **[1]**; they gain electrons **[1]**
      ii) They are oxidised **[1]**; they lose electrons **[1]**
   b) So that the ions can move **[1]**

Pages 56–71 Revise Questions

Page 57 Quick Test

1. A reaction that takes in energy from the surroundings
2. Endothermic – the temperature falls because heat energy is taken from the surroundings
3. a)

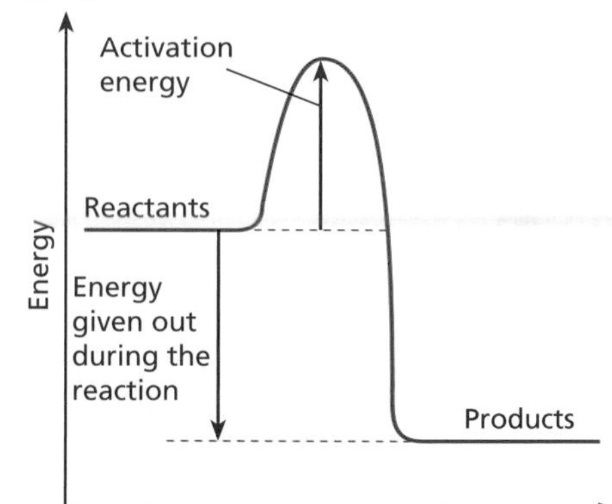

b)

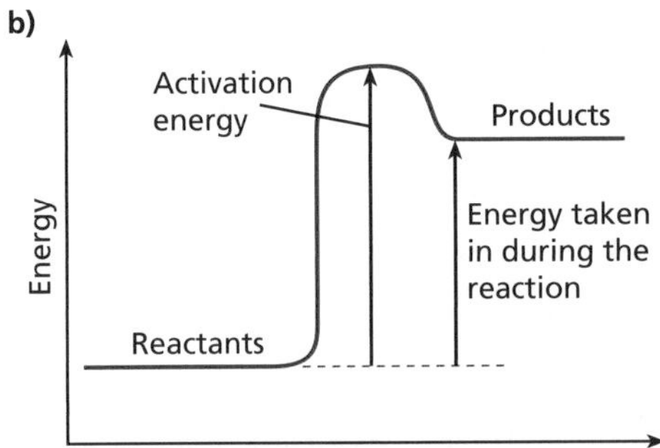

Page 59 Quick Test

1. Exothermic
2. Very efficient; non-polluting; small and lightweight; have no moving parts, so are very reliable
3. a) $2H_2 \rightarrow 4H^+ + 4e^-$
   b) $O_2 + 4H^+ + 4e^- \rightarrow 2H_2O$

Page 61 Quick Test

1. The particles move more quickly. They collide more often, with greater energy, so more collisions are successful.
2. Activation energy
3. A substance that increases the rate of a chemical reaction without being used up or altered in the process

Page 63 Quick Test

1. A reaction that can go forwards or backwards
2. When the rate of the forward reaction is equal to the rate of the backward reaction
3. Using Le Chatelier's Principle

Page 65 Quick Test

1. Molecules that contain only carbon and hydrogen atoms
2. Four
3. C_3H_8
4. C_nH_{2n+2}
5. $CH_4 + 2O_2 \rightarrow CO_2 + 2H_2O$

Page 67 Quick Test

1. They contain at least one C=C bond, so they have the capacity to make further bonds
2. C_2H_4
3. C_5H_{10}
4. Alkenes burn with smokier flames due to incomplete combustion
5. Add bromine water and shake. Butene is an alkene and decolorises the bromine water. Butane does not react.

Page 69 Quick Test

1. 25°C to 50°C
2. $CH_3CH_2CH_2CH_2OH$ or C_4H_9OH
3. $CH_3CH_2CH_2COOH$ or C_3H_7COOH
4. Carbon dioxide and water

Page 71 Quick Test

1. Poly(styrene)
2. 100%
3. A molecule with two hydroxyl, OH, groups
4. Condensation polymerisation
5. Peptide link

Pages 72–77 Review Questions

Page 72 Conservation of Mass

1. a) 12 + (2 × 16) **[1]**; = 44 **[1]**
   b) 44g **[1]**
   c) 88g **[1]**
2. a) Zn **[1]**
   b) $2e^-$ **[1]**; Fe **[1]**
3. a) Mass is conserved / no atoms are gained or lost **[1]**
   b) Solid **[1]**
   c) Carbon dioxide / a gas is made **[1]**; and escapes **[1]**
   d) 10.0 – 5.6 = 4.4g **[1]**

   Remember, total mass before = total mass after.

4. a) (2 × 1) + 16 **[1]**; = 18 **[1]**
   b) 0.5 × 18 **[1]**; = 9g **[1]**
5. a) 24 + (14 + 48) × 2 **[1]**; = 148 **[1]**
   b) 148g **[1]**
6. Oxygen is added **[1]**; and oxygen has mass **[1]**

Page 73 Amount of Substance

1. The number of particles in 1 mole of a substance. **[1]**
2. a) 23g = 1 mole = 6.02×10^{23} atoms **[1]**

   First calculate the number of moles.

   b) 6g = 0.5 moles, so $6.02 \times 10^{23} \times 0.5$ = 3.01×10^{23} atoms **[1]**
3. a) $\frac{39}{39}$ = 1.00mol **[1]**
   b) $\frac{32}{64}$ **[1]**; 0.5mol **[1]**
   c) $\frac{18}{18}$ **[1]**; 1.0mol **[1]**
4. a) $\frac{1.6}{32}$ **[1]**; = 0.05mol **[1]**
   b) 64g **[1]**; 0.05 × 64 = 3.2g **[1]**
5. a) $\frac{0.73}{36.5}$ **[1]**; = 0.02mol **[1]**
   b) $\frac{0.02}{2}$ **[1]**; = 0.01mol **[1]**
   c) 0.01 × 2 **[1]**; = 0.02g **[1]**
6. a) 0.2 × 24 **[1]**; 4.8dm³ **[1]**
   b) $\frac{18}{24}$ **[1]**; = 0.75mol **[1]**

Page 74 Titration

1. $\frac{1}{0.5}$ **[1]**; = 2.00mol/dm³ **[1]**
2. a) $\frac{25}{1000} \times 0.1$ **[1]**; = 0.0025mol **[1]**
   b) 0.0025mol **[1]**

   Look at the overall equation.

   c) $\frac{0.0025 \times 1000}{22.5}$ **[1]**; = 0.11mol/dm³ **[1]**

Page 75 Percentage Yield and Atom Economy

1. $\frac{9}{15}$ **[1]**; = 60% **[1]**
2. a) Copper **[1]**; iron sulfate **[1]**
   b) 56 + 32 + (16 × 4) **[1]**; = 152 **[1]**
   c) $\frac{63.5}{63.5 + 152}$ **[1]**; = 29% **[1]**

Page 76 Reactivity of Metals

1. a) A naturally occurring rock that contains metal, or metal compounds, in sufficient amounts to make it economically worthwhile extracting them **[1]**
   b) Heat it with carbon **[1]**; lead is less reactive than carbon **[1]**

Page 76 The pH Scale and Salts

1. OH^- **[1]**
2. a) Sulfuric acid / hydrochloric acid / nitric acid **[1]**
   b) Split into ions **[1]**
3. 1 **[1]**
4. a) It is one colour in acids and another in alkalis **[1]**
   b) pH probe / pH meter **[1]**
   c) Potassium chloride **[1]**; water **[1]**
   d) 7 **[1]**

Page 77 Electrolysis

1. a) Electrolysis **[1]**
   b) i) Bromine **[1]**

   Negative ions go to the positive electrode.

   ii) Lead **[1]**
   c) So that the ions can move **[1]**
   d) So they do not react with the substances made / electrolyte **[1]**
   e) i) $2e^-$ **[1]**
   ii) Pb **[1]**
2. a) It uses lots of electricity / lots of heat **[1]**
   b) It lowers the melting point of aluminium oxide **[1]**
   c) Aluminium **[1]**
   d) $4e^-$ **[1]**

Pages 78–83 Practice Questions

Page 78 Exothermic and Endothermic Reactions

1. Thermal decomposition **[1]**

Answers

2. Correctly drawn energy profile diagram, showing: reactants **[1]**; products **[1]**; activation energy **[1]**; energy change of reaction **[1]**

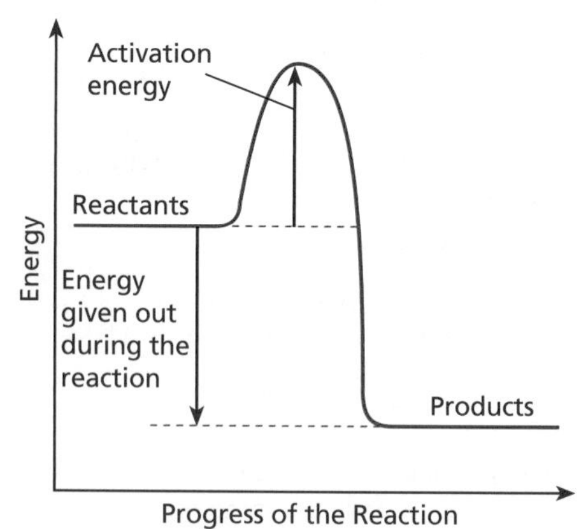

3. **a)** To work out the temperature change **[1]**
   b) To make sure the reactants are completely mixed **[1]**
   c) **Any one of:** same concentration of acid **[1]**; same volume of acid **[1]**

   Control variables must be kept the same.

   d) Acid could damage the student's eyes **[1]**; so eye protection must be worn **[1]**

Page 78 Fuel Cells

1. **Any three of:** only produce water **[1]**; reliable **[1]**; no moving parts so low maintenance **[1]**; small **[1]**; lightweight **[1]**

Page 79 Rate of Reaction

1. Adding a catalyst **[1]**
2. **a)** The particles have to collide **[1]**; the particles have to have enough energy to react / meet the activation energy **[1]**
   b) The particles collide more often **[1]**; the particles have more energy / more have the activation energy **[1]**
3. **a)** Concentration (of acid) **[1]**

   The independent variable is the one that is intentionally changed during the experiment.

   b) Decrease the time **[1]**; because the rate of reaction is faster **[1]**

Page 79 Reversible Reactions

1. **a)** A system where nothing can enter or leave **[1]**
   b) The rate of the forward reaction is equal to the rate of the backward reaction / the relative amounts of all the reacting substances are constant **[1]**

Page 80 Alkanes

1. **a)** A substance that can be burned to release heat **[1]**
   b) $2C_2H_6 + \mathbf{7}O_2 \rightarrow \mathbf{4}CO_2 + \mathbf{7}H_2O$ **[3]**
   (1 mark for each correct number shown in bold)
   c) Soot / particulates / carbon dioxide **[1]**
2. It is the fossilised remains **[1]**; of plankton **[1]**; formed over millions of years **[1]**
3. **a)** Viscous **[1]**
   b) C_2H_4 **[1]**
4. **a)** A molecule that only contains carbon and hydrogen atoms **[1]**; with no double bonds **[1]**
   b) C_nH_{2n+2} **[1]**
   c) C_4H_{10} **[1]**
5. Single covalent **[1]**
6. **a)** **i)** C_3H_8 **[1]**
   ii) C_4H_{10} **[1]**
   b) (Domestic) heating / cooking **[1]**

Page 81 Alkenes

1. C_2H_4 **[1]**
2. **a)** C_3H_6 **[1]**
   b) Orange / brown to colourless **[1]**

Page 82 Organic Compounds

1. Ethyl ethanoate **[1]**
2. Carboxylic acid **[1]**
3. **a)** $C_2H_4 + H_2O$ **[1]** $\rightarrow C_2H_5OH$ **[1]**
   b) The ethene used comes from crude oil **[1]**
   c) 100 **[1]**; there is only one product **[1]**
4. **a)** Amine **[1]**; carboxyl / carboxylic acid **[1]**
   b) Proteins / polypeptides **[1]**
5. **a)** Hydroxyl / OH / alcohol **[1]**
   b) Propanol **[1]**
   c) pH 7 **[1]**
6. **a)** Yeast **[1]**
   b) **i)** Denatures / is destroyed **[1]**
   ii) Becomes inactive / stops working as fast **[1]**
7. **a)** 4 **[1]**
   b) Double helix **[1]**

Page 83 Polymerisation

1. **a)** Poly(ethene) **[1]**
   b) Addition polymerisation **[1]**
   c) 100% **[1]**

   Remember, in an addition reaction all of the reactants are used in the products.

2. **a)** An alcohol with two hydroxyl groups / OH groups **[1]**
   b) Polyester **[1]**
   c) A correctly drawn diagram **[1]**

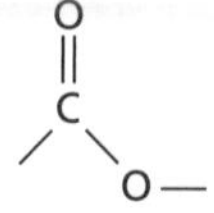

3. Condensation polymerisation **[1]**

Pages 84–97 Revise Questions

Page 85 Quick Test

1. Contains only one type of element or compound
2. **a)** The absorbent paper
   b) The solvent
3. Rapid; very sensitive; accurate; can be used on small samples
4. Solutions containing metal ions

Page 87 Quick Test

1. Green
2. A solid formed when two solutions are mixed
3. Blue
4. Add dilute hydrochloric acid and barium chloride solution; look for a white precipitate of barium sulfate
5. Cream

Page 89 Quick Test

1. Nitrogen
2. Carbon dioxide
3. Mars and Venus
4. Calcium carbonate
5. $6CO_2 + 6H_2O \rightarrow C_6H_{12}O_6 + 6O_2$

Page 91 Quick Test

1. Water vapour; carbon dioxide; methane
2. The Earth would be too cold for water to exist as a liquid, which is needed to support life
3. Combustion of fossil fuels; deforestation
4. Decomposition of rubbish in landfill sites; increase in animal and rice farming
5. Carbon capture and storage / sequestration

Page 93 Quick Test

1. To kill microorganisms
2. As plants grow they absorb (and store) metals such as copper; the plants are then burned and the ash produced contains the metal in relatively high quantities
3. A solution containing bacteria is mixed with a low-grade ore; the bacteria convert the copper into solution (known as a leachate solution) where it can be easily extracted

Page 95 Quick Test

1. A more reactive metal, like magnesium, is placed next to a valuable steel object; the magnesium reacts but the steel does not
2. A layer of aluminum oxide coats the metal and prevents any further reactions taking place
3. To assess the environmental impact a product has over its whole lifetime; this allows people to compare several possible alternative products; to see which one causes the least damage to the environment

Page 97 Quick Test

1. Iron
2. The mixture is cooled and the ammonia liquefies so it can be removed
3. Because plants absorb fertilisers through their roots

4. Nitrogen, phosphorus and potassium
5. Phosphate rock is reacted with sulfuric acid

Pages 98–103 Review Questions

Page 98 Exothermic and Endothermic Reactions

1. Correctly drawn energy profile diagram, showing: reactants **[1]**; products **[1]**; activation energy **[1]**; energy change of reaction **[1]**

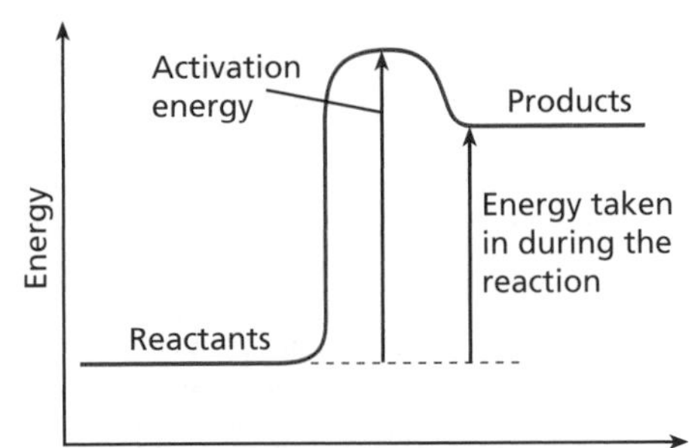

2. The neutralisation of hydrochloric acid by sodium hydroxide solution **[1]**

Page 98 Fuel Cells

1. a) The metal used **[1]**
   b) To make it react faster / increase the surface area **[1]**
   c) **From top to bottom:** 14.5 **[1]**; 33.0 **[1]**; 21.0 **[1]**

Page 99 Rate of Reaction

1. Increasing the size of particles **[1]**
2. The particles are closer together **[1]**; so they collide more often **[1]**
3. a) Iron **[1]**
   b) To reduce energy costs **[1]**
   c) They provide a surface for molecules to attach to **[1]**; this increases their chance of colliding **[1]**
4. a) $\text{mean rate of reaction} = \frac{\text{change in concentration of the reactant (or product)}}{\text{time taken}}$ **[1]**
   b) The concentration of the reactants is greater at the start **[1]**; so particles are colliding more often **[1]**
   c) Draw a tangent to the curve **[1]**; calculate the gradient of the tangent **[1]**

Page 99 Reversible Reactions

1. a) Reversible reaction **[1]**
   b) The rate of the forward reaction is equal to the rate of backward reaction **[1]**

Page 100 Alkanes

1. CH_4 **[1]**

Saturated hydrocarbons only contain single covalent bonds, so each carbon atom is bonded to four other atoms.

2. a) Crude oil is heated **[1]**; the vapour passes up the fractionating column **[1]**; each fraction cools and condenses at a different temperature and is collected **[1]**
   b) Fuel for some cars / lorries / trains **[1]**
3. a) $C_3H_8 + \mathbf{5}O_2 \rightarrow \mathbf{3}CO_2 + \mathbf{4}H_2O$ **[3]**
   (1 mark for each correct number shown in bold)
   b) It is poisonous **[1]**
4. Gases or volatile liquids **[1]**
5. Cracking **[1]**
6. a) Sulfur dioxide **[1]**
   b) Forms acid rain **[1]**; damages statues / buildings / wildlife / trees **[1]**

Page 101 Alkenes

1. a) C_nH_{2n} **[1]**
   b) Contains double bonds **[1]**
   c) C=C **[1]**
   d) C_3H_6 **[1]**
2. Add bromine water **[1]**; shake **[1]**; propene / alkenes decolorise the bromine water **[1]**; propane / alkanes do not react **[1]**
3. Single covalent and double covalent **[1]**
4. a) Nickel **[1]**
   b) C_4H_{10} **[1]**
   c) butene + hydrogen → butane **[1]**
5. a) C_3H_6 **[1]**
   b) Decolorises **[1]**

Page 102 Organic Compounds

1. Alcohol **[1]**
2. a) A molecule with two COOH / carboxyl / carboxylic acid groups **[1]**
   b) Polyester **[1]**
   c) The correct part of the diagram must be circled **[1]**

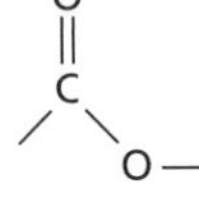

3. Carboxylic acid and amine **[1]**
4. a) Carboxyl / carboxylic acid / COOH **[1]**
   b) Propanoic acid **[1]**
   c) Accept any number less than 7 **[1]**

Page 103 Polymerisation

1. a) C_4H_8 **[1]**
   b) Lots of butene molecules **[1]**; are joined together **[1]**
   c) 1 / 100% **[1]**
2. a) There are only weak forces of attraction **[1]**; between the polymer chains **[1]**
   b) Cooking utensils **[1]** (Accept any other sensible answer)
   c) Crosslinks **[1]**; between the polymer chains **[1]**
3. a) Poly(ethene) **[1]**
   b)

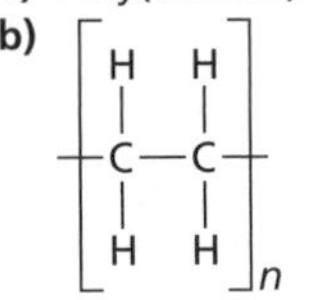

   c) 100% / 1 **[1]**; all the atoms are in the desired product / all the atoms join together **[1]**

Pages 104–109 Practice Questions

Page 104 Chemical Analysis

1. a) Contains only one element or one compound **[1]**
   b) No / it is pure **[1]**; it only has one melting point / it does not have a melting range **[1]**
2. a) Chromatography **[1]**
   b) Fizzy drink 1 **[1]**; it only has one dot / component **[1]**
   c) $\frac{6.0}{12.0}$ **[1]**; = 0.5 **[1]**
   d) Fizzy drink 1 **[1]**; it also contains a component with a R_f value of 0.5 / the dot travels the same distance **[1]**

Page 104 Identifying Substances

1. Lilac **[1]**
2. Green **[1]**
3. a) Add nitric acid **[1]**; add silver nitrate solution **[1]**; if the sample contains chloride ions, you would expect to see a white precipitate **[1]**
   b) The sodium ions produce a yellow flame **[1]**; the potassium ions produce a lilac flame **[1]**; the yellow flame may mask the lilac flame / is stronger than the lilac flame **[1]**
4. a) Add dilute hydrochloric acid then barium chloride solution **[1]**
   b) White precipitate **[1]**
5. a) i) It has a high melting point / does not melt **[1]**
   ii) To ensure it is clean **[1]**
   b) Green **[1]**
6. Add dilute acid **[1]**; collect gas and bubble it through limewater **[1]**; if carbonate ions are present the limewater will go cloudy / milky **[1]**

Page 106 The Earth's Atmosphere

1. 4.6 billion **[1]**
2. Nitrogen (78%) **[1]**; oxygen (21%) **[1]**

The main gas in today's atmosphere is nitrogen.

3. a) Carbon dioxide **[1]**
   b) Mars **[1]**; Venus **[1]**
   c) They photosynthesise / give out oxygen **[1]**; so the levels of oxygen have risen **[1]**
4. a) carbon dioxide + water → glucose + oxygen **[1]**
   b) 6 **[1]**; 6 **[1]**
   c) Plants take in carbon dioxide for photosynthesis **[1]**; reducing the levels in the atmosphere **[1]**

Page 106 Greenhouse Gases

1. Water vapour / carbon dioxide **[1]**
2. More animal farming **[1]**
3. a) Too complicated / too many factors **[1]**

b) They are biased [1]

c) People who work for the government / universities / are well qualified [1]

4. a) Flooding / coastal erosion [1]

b) Some areas become too hot or too cold **[1]**; for the organisms that currently live there **[1]**

c) Carbon capture and storage / sequestration [1]

d) It stops carbon dioxide going into the atmosphere [1]

5. a) The total amount of carbon dioxide (and other greenhouse gases) **[1]**; given out in the production, use and disposal of the product / over the product's lifetime **[1]**

b) Leads to no overall increase in the amount of carbon dioxide in the atmosphere [1]

c) People may be encouraged to buy it **[1]** (Accept any other sensible answer)

d) They might encourage companies and individuals to choose other options **[1]**; leading to less carbon dioxide being produced **[1]**

6. a) Some areas will have more **[1]**; and some areas will have less **[1]**

b) Some areas will be too hot to grow food crops **[1]**; and some areas will be too dry **[1]**

Page 108 Earth's Resources

1. a) Water with low levels of salts and microorganisms [1]

b) Potable [1]

c) To remove solids [1]

d) To kill microorganisms [1]

2. Ozone [1]

3. a) A naturally occurring rock that contains metal, or metal compounds, in sufficient amounts to make it economically worthwhile extracting them [1]

b) The ore is heated **[1]**; with carbon (which is more reactive than copper) **[1]**

c) Bacteria **[1]**; are used to remove copper from a leachate solution **[1]**

Page 108 Using Resources

1. a) Oxygen / air **[1]**; and water **[1]**

b) Galvanising [1]

c) Zinc is more reactive so it reacts first **[1]**; preventing the iron from rusting / sacrificial protection **[1]**

2. The aluminium reacts forming a layer of aluminium oxide **[1]**; which protects the metal / stops further reactions **[1]**

3. Magnesium / zinc is placed in contact with the steel **[1]**; the magnesium / zinc is more reactive and reacts first **[1]**; so that the steel does not react **[1]**

Page 109 The Haber Process

1. a) nitrogen + hydrogen $\rightleftharpoons$ ammonia [1]

b) (Fractional distillation of liquid) air [1]

c) 200atm [1]

d) A high pressure **[1]**; increases the yield of ammonia **[1]**

e) The mixture is cooled **[1]**; and the ammonia liquefies / condenses **[1]**

2. a) To replace nutrients in the soil, so that plants grow well [1]

b) So they can be taken up through the roots of plants [1]

c) Nitrogen, phosphorus and potassium [1]

Pages 110–115 Review Questions

Page 110 Chemical Analysis

1. a) i) Absorbent paper [1]

ii) Solvent / water [1]

b) It does not run / ink would dissolve and affect the results [1]

c) $\frac{4.5}{10.0}$ **[1]**; = 0.45 **[1]**

d) It produces two spots / it has two components [1]

2. a) Fast / sensitive / can be used on small samples [1]

b) The metal ions present **[1]**; and the concentration of the metal ions present **[1]**

Page 111 Identifying Substances

1. a) Solid [1]

b) Gas [1]

c) Bubble the gas through limewater **[1]**; it will go cloudy / white if the gas is carbon dioxide **[1]**

2. Yellow [1]

3. a) They both contain sodium / the same metal **[1]**; and will give a yellow flame **[1]**

b) A white precipitate is produced from the sodium chloride **[1]**; a cream precipitate is produced from the sodium bromide **[1]**

4. a) Dilute hydrochloric acid **[1]**; and barium chloride **[1]**

b) A white precipitate **[1]**; of barium sulfate **[1]**

5. Blue [1]

6. A glowing splint placed in the gas **[1]**; will relight if it is oxygen **[1]**

7. Hold damp litmus paper in the gas **[1]**; it will be bleached if the chlorine is gas **[1]**

Page 112 The Earth's Atmosphere

1. a) 2.7 billion [1]

b) carbon dioxide **[1]**; oxygen **[1]**

2. a) By volcanic activity [1]

b) Carbon dioxide [1]

c) By bacteria [1]

d) Carbon dioxide levels went down **[1]**; as carbon was locked-up in rocks **[1]**

3. Mars [1]

4. a) Coal contains carbon **[1]**; when it is burned it makes carbon dioxide **[1]**

b) Plant deposits / remains **[1]**; buried / compressed **[1]**; over millions of years **[1]**

5. a) Insoluble carbonates **[1]**; soluble hydrogen carbonates **[1]**

b) Algae take carbon dioxide out of the atmosphere for photosynthesis [1]

Page 113 Greenhouse Gases

1. a) CO_2 [1]

b) They absorb / trap **[1]**; infrared radiation **[1]**

2. Deforestation [1]

3. a) Carbon dioxide [1]

b) Trees photosynthesise **[1]**; taking in carbon dioxide from the atmosphere, which reduces the level of carbon dioxide **[1]**

Page 114 Earth's Resources

1. Ultraviolet radiation [1]

2. a) It is a good electrical conductor **[1]**; it bends easily **[1]**

b) It is a good thermal conductor [1]

c) Copper is less reactive than carbon [1]

d) Plants **[1]**; grow and take up the metal **[1]**; which is collected by burning the plants **[1]**

3. Digging up large amounts of rock **[1]**; moving large amounts of rock **[1]**; and disposing of large amounts of waste material all have a negative impact on the environment **[1]**

Page 114 Using Resources

1. a) It stops oxygen / air **[1]**; and water from coming into contact with the iron **[1]**

b) It will rust **[1]**; because oxygen / air and water can reach the iron / plastic will not provide sacrificial protection **[1]**

Page 115 The Haber Process

1. a) $N_2 + \mathbf{3}H_2 \rightleftharpoons \mathbf{2}NH_3$ [2]
(1 mark for each correct number shown in bold)

b) Reversible reaction [1]

c) From methane / steam [1]

d) Iron [1]

e) To speed up the reaction / it allows a lower temperature to be used [1]

f) i) Faster rate **[1]**; lower yield **[1]**

ii) Slower rate **[1]**; higher yield **[1]**

iii) It is a good compromise / the best option **[1]**; to produce a reasonable yield at a reasonable rate **[1]**

2. Fertiliser B [1]

Pages 116–127 Mixed Exam-Style Questions

1. a) 2,8,8,1 [1]

b) It has one electron **[1]**; in its outer shell **[1]**

c) A correctly drawn diagram [1]

d) Both have one electron in their outer shell [1]

2. A substance that speeds up a reaction [1]; but is not used up **[1]**
3. An atom that has gained electrons **[1]**; or lost electrons **[1]**
4. 3 protons **[1]**; 2 electrons **[1]**; 4 neutrons **[1]**
5. a) The only product is water [1]
   b) $2H_2 + O_2 \rightarrow 2H_2O$ **[2]** (**1** mark for correct reactants and products; **1** mark for correct balancing)

   Add the two equations together.

6. a) Tin **[1]**; carbon dioxide **[1]**
   b) Tin oxide is reduced **[1]**; carbon is oxidised **[1]**
7. Sodium sulfate [1]
8. **Properties ticked should be:** very strong **[1]**; good electrical conductor **[1]**; good thermal conductor **[1]**; and nearly transparent **[1]**
9. a) It contains lots of strong bonds **[1]**; that require lots of energy to overcome them **[1]**
   b) No, because the ions cannot move [1]
   c) Yes, because the ions can move [1]
   d) Yes, because the ions can move [1]
10. a) 20 [1]
    b) 21 [1]
    c) 20 [1]
11. a) 14 + (3 × 1) **[1]**; = 17 **[1]**
    b) 17g [1]
    c) 0.5 × 17 = 8.5g [1]
12. a) $\mathbf{2}Ca + O_2 \rightarrow \mathbf{2}CaO$ [2] (1 mark for each correct number shown in bold)
    b) $\mathbf{2}Na + Br_2 \rightarrow \mathbf{2}NaBr$ [2] (1 mark for each correct number shown in bold)
    c) $H_2 + Br_2 \rightarrow \mathbf{2}HBr$ (1 mark for each correct number shown in bold) [1]
    d) $\mathbf{3}H_2 + N_2 \rightarrow \mathbf{2}NH_3$ [2] (1 mark for each correct number shown in bold)
    e) $\mathbf{2}K + I_2 \rightarrow \mathbf{2}KI$ [2] (1 mark for each correct number shown in bold)
13. Distillation [1]
14. a) 2,1 [1]
    b) It has one electron in its outer shell [1]
15. a) Correctly drawn diagrams showing particles in a solid **[1]**; liquid **[1]**; and gas **[1]**

    Solid Liquid Gas

    b) Quickly **[1]**; in all directions **[1]**
    c) liquid **[1]**; gaseous **[1]**
    d) Water has covalent bonds **[1]**; and a simple molecular structure **[1]**; little energy is required to overcome the forces of attraction between molecules **[1]**; iron has metallic bonds **[1]**; and a metallic structure **[1]**; lots of energy is required to overcome the strong metallic bonds / bonds between the metal ions and the delocalised electrons **[1]**
16. a) Gained **[1]**; 1 electron **[1]**

    Remember that electrons have a negative charge.

    b) Lost **[1]**; 2 electrons **[1]**
17. a) Ag [1]
    b) Metallic [1]
    c) It has a regular structure **[1]**; so layers can slide over each other **[1]**
    d) A mixture that contains at least one metal [1]
18. a) $Mg + \mathbf{2}HCl \rightarrow MgCl_2 + H_2$ [1]
    b) $\mathbf{2}HgO \rightarrow \mathbf{2}Hg + O_2$ [2] (1 mark for each correct number shown in bold)
19. a) C_5H_{10} [1]
    b) Lots of **[1]**; pentene molecules **[1]**; are joined together **[1]**
    c) 1 / 100% [1]
20. Single covalent [1]
21. a) i) Group 2 [1]
       ii) Group 6 [1]
       iii) A diagram showing the correctly drawn outer electrons for a magnesium ion **[1]**; and oxide ion **[1]**

    $[Mg]^{2+}$ $[O]^{2-}$

    b) Giant ionic [1]
    c) Ions cannot move in a solid **[1]**; but they can move when the substance is molten **[1]**
22. a) Covalent [1]
    b) A correctly drawn diagram [1]
23. a) Fractional distillation [1]
    b) $\mathbf{2}C_2H_6 + \mathbf{7}O_2 \rightarrow \mathbf{4}CO_2 + \mathbf{7}H_2O$ [3] (1 mark for each correct number shown in bold)
    c) It is a poisonous gas [1]
24. a) $\frac{25}{1000} \times 0.15$ **[1]**; 0.00375mol **[1]**
    b) 0.00375mol [1]
    c) $\frac{0.00375 \times 1000}{28}$ **[1]**; = $0.134mol/dm^3$**[1]**
25. a) C_nH_{2n} [1]
    b) C=C [1]
    c) C_4H_8 [1]
26. $\frac{8.5}{15} \times 100$ **[1]**; = 56.7% **[1]**

    percentage yield (%) $= \frac{\text{actual yield}}{\text{theoretical yield}} \times 100$

27. a) $\frac{39}{39}$ = 1 mole = 6.02×10^{23} atoms [1]
    b) $\frac{16}{32}$ = 0.5 mole = 3.01×10^{23} atoms [1]
28. C_3H_8 [1]
29. a) i) Bromine [1]
       ii) Lead [1]
    b) i) Br_2 [1]
       ii) $2e^-$ [1]
30. a) Molecules with two COOH / carboxyl groups / carboxylic acid groups [1]
    b) Polyester [1]
    c) Condensation **[1]**; polymerisation **[1]**
    d) A correct repeat unit drawn [1]
31. a) To replace nutrients in the soil so that plants grow well [1]
    b) So they can be absorbed through the roots of the plant [1]
    c) Potassium [1]
    d) Fertiliser C [1]
32. a) Carboxyl / carboxylic acid / COOH [1]
    b) Butanoic acid [1]
    c) Any pH less than 7 [1]
    d) H^+ [1]
33. a) Particles must collide **[1]**; and have enough energy to react **[1]**
    b) Particles are closer together **[1]**; so they collide more often **[1]**
34. a) Each lead ion gains two electrons **[1]**; to form lead atoms **[1]**
    b) The iodide loses electrons **[1]**; so it is oxidised **[1]**

    Remember: OIL RIG

    c) So the ions can move [1]
35. a) Magnesium is placed on the surface of the iron / steel **[1]**; magnesium is more reactive than iron / steel **[1]**; so the magnesium reacts but the iron / steel does not **[1]**
    b) Copper is less reactive than iron **[1]**; so the iron / steel would still react **[1]**
36. Brown [1]
37. a) Steel coated in zinc [1]
    b) Zinc is more reactive **[1]**; so the zinc reacts but steel does not **[1]**
38. a) A naturally occurring rock that contains metal or metal compounds in sufficient amounts to make it economically worthwhile extracting them [1]
    b) The ore is heated **[1]**; with carbon **[1]**
    c) Lead is less reactive than carbon [1]
    d) When plants are used to take in metals as they grow [1]
    e) Aluminium reacts with oxygen / forms a layer of aluminium oxide **[1]**; which stops any further reactions **[1]**
39. Red [1]

Answers

40. a) Ammonia [1]
    b) Reversible reaction [1]
    c) 450°C [1]
    d) It would give a faster rate of reaction **[1]**; but a lower yield **[1]**
    e) They are recycled [1]
41. Increasing the surface area of solids [1]
42. a) A substance that can be burned to release heat [1]
    b) Fractional distillation [1]
    c) A correctly drawn diagram showing the reactants **[1]**; products **[1]**; activation energy **[1]**; and the energy change of the reaction **[1]**

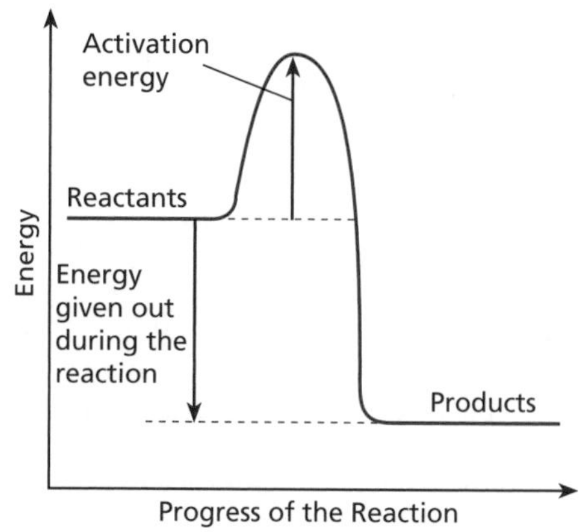

43. a) So it does not run / because ink would dissolve and affect the results [1]
    b) $\frac{8}{10}$ **[1]**; = 0.8 **[1]**
    c) There are two dots / two components [1]
44. a) Carry out a flame test **[1]**; there will be a yellow flame if sodium ions are present **[1]**
    b) i) Nitric acid **[1]**; and barium chloride **[1]**
       ii) A white precipitate **[1]**; of barium sulfate **[1]**
45. a) Solid [1]
    b) Gas [1]
    c) A strong acid produces H^+ ions **[1]**; and is fully ionised **[1]**
    d) The gas carbon dioxide is made **[1]**; and escapes **[1]**

Notes

Notes

Notes

Notes

Key

relative atomic mass **atomic symbol** name atomic (proton) number

1	2											3	4	5	6	7	0
							1 **H** hydrogen 1										4 **He** helium 2
7 **Li** lithium 3	9 **Be** beryllium 4											11 **B** boron 5	12 **C** carbon 6	14 **N** nitrogen 7	16 **O** oxygen 8	19 **F** fluorine 9	20 **Ne** neon 10
23 **Na** sodium 11	24 **Mg** magnesium 12											27 **Al** aluminum 13	28 **Si** silicon 14	31 **P** phosphorus 15	32 **S** sulfur 16	35.5 **Cl** chlorine 17	40 **Ar** argon 18
39 **K** potassium 19	40 **Ca** calcium 20	45 **Sc** scandium 21	48 **Ti** titanium 22	51 **V** vanadium 23	52 **Cr** chromium 24	55 **Mn** manganese 25	56 **Fe** iron 26	59 **Co** cobalt 27	59 **Ni** nickel 28	63.5 **Cu** copper 29	65 **Zn** zinc 30	70 **Ga** gallium 31	73 **Ge** germanium 32	75 **As** arsenic 33	79 **Se** selenium 34	80 **Br** bromine 35	84 **Kr** krypton 36
85 **Rb** rubidium 37	88 **Sr** strontium 38	89 **Y** yttrium 39	91 **Zr** zirconium 40	93 **Nb** niobium 41	96 **Mo** molybdenum 42	[98] **Tc** technetium 43	101 **Ru** ruthenium 44	103 **Rh** rhodium 45	106 **Pd** palladium 46	108 **Ag** silver 47	112 **Cd** cadmium 48	115 **In** indium 49	119 **Sn** tin 50	122 **Sb** antimony 51	128 **Te** tellurium 52	127 **I** iodine 53	131 **Xe** xenon 54
133 **Cs** cesium 55	137 **Ba** barium 56	139 **La*** lanthanum 57	178 **Hf** hafnium 72	181 **Ta** tantalum 73	184 **W** tungsten 74	186 **Re** rhenium 75	190 **Os** osmium 76	192 **Ir** iridium 77	195 **Pt** platinum 78	197 **Au** gold 79	201 **Hg** mercury 80	204 **Tl** thallium 81	207 **Pb** lead 82	209 **Bi** bismuth 83	[209] **Po** polonium 84	[210] **At** astatine 85	[222] **Rn** radon 86
[223] **Fr** francium 87	[226] **Ra** radium 88	[227] **Ac*** actinium 89	[261] **Rf** rutherfordium 104	[262] **Db** dubnium 105	[266] **Sg** seaborgium 106	[264] **Bh** bohrium 107	[277] **Hs** hassium 108	[268] **Mt** meitnerium 109	[271] **Ds** darmstadtium 110	[272] **Rg** roentgenium 111	Elements with atomic numbers 112–116 have been reported but not fully authenticated						

* The Lanthanides (atomic numbers 58–71) and the Actinides (atomic numbers 90–103) have been omitted.
Relative atomic masses for **Cu** and **Cl** have not been rounded to the nearest whole number.

Glossary and Index

A

Activation energy the minimum amount of energy required for a reaction to take place 57

Addition polymerisation when many monomers join together to form a large molecule 70–71

Addition reaction a reaction in which the reactants join together to form one product 37, 67

Agriculture farming 92

Algae small green aquatic plants 89

Alkali metals elements in Group 1 of the periodic table 12–13

Alkanes saturated hydrocarbons with the general formula C_nH_{2n+2} 64–65

Alkenes unsaturated hydrocarbons with the general formula C_nH_{2n} 66–67, 70–71

Alloy a mixture that contains at least one metal, e.g. steel 18–19

HT **Amino acid** a simple organic compound containing both a carboxyl (–COOH) and an amino (–NH_2) group 69, 71

Anode a positively charged electrode 42–43

Aqueous dissolved in water 15, 40–41

Atom the smallest part of an element that can exist 8–9

Atom economy a measure of the amount of reactant converted to a useful product,

i.e. $\frac{\text{relative formula mass of the desired product}}{\text{sum of the relative formula mass of all the products/reactants}}$; it can be stated as a number or a percentage 36–37

Atomic number the number of protons in the nucleus of an atom 11

HT **Avogadro constant** the number of particles in one mole of any substance, i.e. six hundred thousand billion billion or 6.02×10^{23} 32–33

B

Biased has a prejudice; is not a balanced view 90

Billion 1000 000 000 or one thousand million 88

HT **Bioleaching** the use of bacteria to extract metals from low-grade ores 93

Boiling point the specific temperature at which a pure substance changes state from liquid to gas and from gas to liquid 14–15

Brittle breaks easily when hit 19

C

Carbon footprint the total amount of carbon dioxide and other greenhouse gases that are emitted over the full life cycle of a product, a service or an event 91

Carboxyl the -COOH group found in carboxylic acids 68–69

Catalyst a substance that speeds up the rate of a chemical reaction, but is not used up in the reaction itself 13, 57, 62, 96

Cathode a negatively charged electrode 42–43

Cell (electrical) contains chemicals that react together to release electricity 59

Cellulose a form of carbohydrate 69

Chromatography a separation technique used to separate the coloured components of mixtures 9, 84–85

Closed system where nothing can enter or leave 63

Combustion burning 65

Composite material a material consisting of two materials with different properties, which are combined together to produce an improved material with its own properties that are better than either of the individual components' properties 95

Compound a substance containing atoms of two or more elements, which are chemically combined in fixed proportions 8–9

HT **Concentration** the amount of substance in a given volume, normally measured in units of mol/dm^3 34–35, 60–61

HT **Condensation polymerisation** when many small monomer molecules join together to form large polymer molecules and small molecules such as water, H_2O 71

Conservation of mass the total mass of the products of a chemical reaction is always equal to the total mass of the reactants 30–31

Corrode / corrosion when a metal reacts with oxygen and water, e.g. the rusting of iron 19, 94

Covalent bond a shared pair of electrons between atoms in a molecule 20–21

Cracking the process by which longer-chain hydrocarbons can be broken down into shorter, more useful hydrocarbons 66

Crude oil a fossil fuel; a mixture consisting mainly of alkanes 64–65

Cryolite a compound of aluminium 43

Crystallisation technique used to obtain a soluble solid from a solution 9

D

Delocalised not bound to one atom 18–19, 21

Desalinated salt water that has had all the dissolved salts removed to make pure water 92

Diamond a form of carbon, with a giant covalent structure, that is very hard 21

HT **Dicarboxylic acid** a molecule with two carboxyl, -COOH, groups 71

HT **Diol** a molecule with two hydroxyl, -OH, groups 71

Displace to take the place of (another element, e.g. in a reaction) 13

Displacement reaction a reaction in which a more reactive metal displaces a less reactive metal from a solution of its salt 39

Dissociate to split up into ions 40–41

DNA a very large molecule with a double helix structure, which stores and transmits instructions for the development of living organisms and some viruses 69

Ductile can be drawn into a wire 18

E

Electrode an electrical conductor used in a cell 42–43

Electrolysis the process by which an ionic compound is broken down into its elements using an electrical current 42–43

Electrolyte the substance being broken down during electrolysis 42–43

Electron a subatomic particle with a relative mass that is very small and relative charge of –1 10–11

Electron configuration represents how the electrons are arranged in shells around the nucleus of an atom 11

Electrostatic a force of attraction between oppositely charged species 16–19

Element a substance made of only one type of atom 8–9

Endothermic reaction a reaction that takes in energy from the surroundings 56–59

Energy level diagram (reaction profile) a diagram showing the relative energies of the reactants and the products of a reaction, the activation energy and the overall energy change of the reaction 57

Equation a scientific statement that uses chemical names or symbols to sum up what happens in a chemical reaction 8–9

Equilibrium when the rate of forward reaction is equal to the rate of backwards reaction 63

Exothermic reaction a reaction that gives out energy to the surroundings 56–59, 96

Extraction the process of obtaining / taking out / removing, e.g. obtaining a metal from an ore or compound 39

F

Fermentation a chemical reaction in which alcohol and carbon dioxide are produced from glucose 37, 68–69

Filtration separation techniques used to separate insoluble solids from soluble solids 9

Flame emission spectroscopy an instrumental method of analysis, which is used to identify the metal ions in a solution and measure their concentration 85

Flame tests a way of identifying metal ions from the colour produced when the metal compound is heated in a flame 86

Formulation a mixture that has been carefully designed to have specific properties 84

Fractional distillation a separation technique used to separate mixtures which contain components with similar boiling points 9, 64–65

Fresh water water that contains low levels of dissolved salts 92

Fuel cell a very efficient way of producing electrical energy in which a fuel is oxidised electrochemically to produce a potential difference or voltage 59

Fullerene a form of carbon in which the carbon atoms are joined together to form hollow structures, e.g. tubes, balls and cages 22

G

Galvanising a method of protecting iron or steel objects by covering them in a coating of zinc 94

HT **Gradient** $\text{gradient} = \dfrac{\text{difference in the } y\text{-axis value}}{\text{difference in the } x\text{-axis value}}$ 61

Graphene a form of carbon; a single layer of graphite, just one atom thick 22

Graphite a form of carbon, with a giant covalent structure, that conducts electricity 21

Greenhouse gases gases (e.g. carbon dioxide, water vapour or methane) that, when present in the atmosphere, absorb the outgoing infra-red radiation, increasing the Earth's temperature 90–91

H

Haber process an industrial process used to manufacture ammonia, NH_3 96

HT **Half equation** used to show what happens to one of the reactants in a chemical reaction 30

Halide a halogen ion, e.g. fluoride, chloride, bromide or iodide 87

Halogens elements in Group 7 of the periodic table 13

Homologous series members of the same family of compounds; they have the same functional group and each one differs from the previous one by the addition of CH_2 68

Hydration a reaction in which H_2O is added to a substance 37

Hydrocarbon a molecule that contains only carbon and hydrogen atoms 64–65

Hydroxyl the -OH group found in alcohols 68

I

Indicator a chemical that is one colour in an acid and another colour in an alkali 34–35, 40-41

Infrared a type of radiation that is part of the electromagnetic spectrum 90

Insoluble a substance that does not dissolve in a solvent 87

Intermolecular between molecules 20–21

Ion atoms that have gained or lost electrons and now have an overall charge 11, 16–17

Ionic bond the force of attraction between positive and negative ions 16–17

HT **Ionic equation** a simplified version of a chemical equation, which just shows the species that are involved in the reaction 30

HT **Ionised** split up into ions 41

Isotope atoms of the same element that have the same atomic number but a different mass number 11

L

HT **Le Chatelier's Principle** a rule that can predict the effect of changing conditions on a system that is in equilibrium 63

Life cycle assessment (LCA) used to assess the environmental impact a product has over its whole lifetime 95

HT **Limiting reactant** the reactant that is completely used up in a reaction; it stops the reaction from going any further and any further products being produced 33

M

Malleable can be hammered into shape 18

Mass number the total number of protons and neutrons in the nucleus of an atom 11

Matter describes the material that everything is made up of 14–15

Melting point the specific temperature at which a pure substance changes state, from solid to liquid and from liquid to solid 14–15

Mendeleev the Russian chemist who devised the periodic table 12

Metallic bond the strong attraction between metal ions and delocalised electrons 18–19

Mixture a combination of two or more elements or compounds, which are not chemically combined together 8–9

Mobile phase in chromatography, the phase that does move (e.g. the solvent) 84–85

HT **Mole (mol)** the unit for measuring the amount of substance; one mole of any substance contains the same number of particles; the mass of one mole of a substance is equal to the relative formula mass in grams 32–33

Molten liquefied by heat 17

Monomer a small molecule that can join together to form a polymer 70–71

N

Nanoparticles very small particles, which are 1–100nm in size (just a few hundred atoms) 23

Neutralisation occurs when an acid reacts with exactly the right amount of base 40–41

Neutron a subatomic particle with a relative mass of 1 and no charge 10–11

Noble gases unreactive non-metal elements that have a full outer shell of electrons and are found in Group 0 of the periodic table 12

NPK fertilisers fertilisers that contain nitrogen, phosphorus and potassium to help plants grow better 97

O

Oxidation when a species loses electrons 38–39

P

Particle a small piece of matter 14–15

Percentage yield a calculation used to compare the actual yield of a product obtained from a reaction with the maximum theoretical yield 36–37

Phosphate rock a rock that contains high level of phosphorus compounds 97

Photosynthesis a chemical reaction carried out by green plants in which carbon dioxide and water react to produce glucose and oxygen 89

HT **Phytomining** the use of plants to extract metals from low-grade ores 93

HT **Polyester** a type of condensation polymer **71**

Polymer large molecules formed when lots of smaller monomer molecules join together **70–71**

Potable water that is safe to drink **92–93**

Precipitate a solid formed when two solutions are combined **87**

Products the substances produced in a chemical reaction **8–9**

Proton a subatomic particle with a relative mass of 1 and relative charge of +1 **10–11**

Pure a substance that contains only one type of element or compound **18, 84**

R

Rate of reaction a measure of how quickly a reactant is used up or a product is made **60–61**

Reactants the substances that react together in a chemical reaction **8–9**

Reaction profile (energy level diagram) a diagram showing the relative energies of the reactants and the products or a reaction, the activation energy and the overall energy change of the reaction **57**

Reactivity series a list in which elements are placed in order of how reactive they are **38–39**

Reduction when a species gains electrons **38–39**

Relative atomic mass (A_r) the ratio of the average mass per atom of the element to one-twelfth the mass of an atom of carbon-12 **31**

Relative formula mass (M_r) the sum of the relative atomic masses of all the atoms shown in the formula **31**

Reversible reaction a reaction that can go forwards or backwards **56–57, 62–63, 96**

Rusting when iron or steel objects corrode **94**

S

Sacrificial protection a method of protecting iron or steel objects by placing a more reactive metal, such as magnesium, in contact with the object – the magnesium reacts but the object does not **94**

Saturated a hydrocarbon molecule that only contains single carbon–carbon (C–C) bonds **64–65**

Silicon dioxide (silica) (SiO_2) a compound found in sand **21**

Simple distillation technique used to separate a substance from a mixture due to a difference in the boiling points of the components in the mixture **9**

HT **Solute** a substance that dissolves in a solvent to form a mixture called a solution **34–35**

HT **Species** the different atoms, molecules or ions that are involved in a reaction **30–31**

Stationary phase in chromatography, the phase that does not move (e.g. the absorbent paper) **84–85**

HT **Strong acid** an acid, such as hydrochloric acid, nitric acid or sulfuric acid, that is completely ionised in water **41**

Sustainable development development that meets the needs of the current generation without compromising the ability of future generations to meet their own needs **37, 92–93**

T

HT **Tangent** a straight line that touches a curve at one point **61**

HT **Titration** an accurate technique that can be used to find out how much of an acid is needed to neutralise an alkali **34–35**

Transition metals metals that have typical metal properties and are found in the middle of the periodic table **13**

Transparent see-through **22**

U

Unsaturated a hydrocarbon molecule with at least one double carbon–carbon (C=C) bond **66–67, 70–71**

V

Viscous sticky **64**

W

HT **Weak acid** an acid, such as ethanoic acid, citric acid or carbonic acid, that is only partially ionised in water **41**